Contents

D1152492

Getting the most from this book 4
About this book 5

Content Guidance

Physical quantities 6
Scalars and vectors 9
Linear motion 12
Projectiles 20
Newton's laws of motion 22
Principle of moments 25
Work done, potential and kinetic energy 29
Deformation of solids 34
Electricity 37
Circuits 39
Current–voltage characteristics 43
Ohm's law 45
Resistivity 47
Internal resistance and electromotive force 50
Potential divider 52

Questions & Answers

Self-assessment test 1 59
Self-assessment test 2 70

Knowledge check answers 78
Index 79

LIBRARIES NI
WITHDRAWN FROM STOCK

Getting the most from this book

Examiner tips

Advice from the examiner on key points in the text to help you learn and recall unit content, avoid pitfalls, and polish your exam technique in order to boost your grade.

Knowledge check

Rapid-fire questions throughout the Content Guidance section to check your understanding.

Knowledge check answers

1 Turn to the back of the book for the Knowledge check answers.

Summary

Summaries

- Each core topic is rounded off by a bullet-list summary for quick-check reference of what you need to know.

Questions & Answers

Exam-style questions

Examiner comments on the questions
Tips on what you need to do to gain full marks, indicated by the icon e.

Sample student answers
Practise the questions, then look at the student answers that follow each question.

Self-assessment test 1 Question 1

Self-assessment test 1

Question 1

(a) Show that the base units of potential energy and kinetic energy are the same. (4 marks)

e Be clear of the difference between the SI unit for the quantity and the base units. The SI unit for any form of energy is the joule. Note that the SI unit of mass is the kilogram and not the gram.

(b) A mass m is attached to one end of a suspended spiral spring. This produces a force F in the spring. The mass is then pulled down and released, causing it to oscillate. Equation 1 represents the relationship for the periodic time T of mass–spring oscillation.

$T = 2\pi\sqrt{\frac{my}{F}}$ **Equation 1**

Determine the base units of the term y. (2 marks)

Answer to Question 1

(a) potential energy = mgh ✔
where g is the acceleration due to gravity.
units of mass = kg
units of acceleration = $m\,s^{-2}$
units of height = m
base units of potential energy = (kg) × ($m\,s^{-2}$) × (m) = $kg\,m^2\,s^{-2}$ ✔
kinetic energy = $\frac{1}{2}mv^2$ ✔
units of mass = kg
units of velocity = $m\,s^{-1}$
base units of kinetic energy = (kg) × $(m\,s^{-1})^2$ = (kg) × ($m\,s^{-1}$) × ($m\,s^{-1}$) = $kg\,m^2\,s^{-2}$ ✔
Both forms of energy have the base units $kg\,m^2\,s^{-2}$.

e Make sure you show all stages of the solution to convince the examiner that your answer is not a bluff.

(b) base units of periodic time T = s
base units of mass m = kg
base units of force F = $kg\,m\,s^{-2}$ ✔

e It is often simpler to remove a square root by squaring both sides of the equation.

Unit 1: Forces, Energy and Electricity 59

Examiner commentary on sample student answers
Find out how many marks each answer would be awarded in the exam and then read the examiner comments (preceded by the icon e) following each student answer.

About this book

This book covers the CCEA specification for AS Physics **Unit 1: Forces, Energy and Electricity**. It has two sections:

- **Content Guidance** — this covers AS Unit 1. It does not have the detail of a textbook but it offers guidance on the main areas of the content and includes worked examples. These examples illustrate the types of question that you are likely to come across in the examination. The examiner tips will help you to understand the physics and give you guidance on the core aspects of the subject. They also show how to approach revision and improve your exam technique.
- **Questions and Answers** — this section comprises two self-assessment tests. Answers are provided and there is an indication of the specific points for which marks are awarded.

The specification

The CCEA specification can be obtained from the CCEA website at **www.rewardinglearning.org.uk**. It describes the physics that is required for the unit assessments, and the format of the assessments.

Content Guidance

Physical quantities

Physics is often referred to as the quantitative science. When we measure an amount of any physical quantity, we do not do so in absolute terms, but we compare its size with internationally accepted amounts of a few basic quantities.

All the quantities have magnitude, but what differentiates one from another is its unit.

- A mass of 5 kg can be thought of as 5 × 1 kg.
- A length of 5 m can be thought of as 5 × 1 m.

Both examples have the same magnitude but their difference is conveyed by their specific units.

SI base units

It is essential to have a coherent, internationally accepted system of units.

The International System of Units (SI) has seven **base units**, each exclusive to a fundamental physical quantity (Table 1).

Table 1 Base quantities and SI units

Base quantity	Name of SI base unit	Symbol for unit
Length	metre	m
Mass	kilogram	kg
Time	second	s
Electric current	ampere	A
Thermodynamic temperature	kelvin	K
Luminous intensity*	candela	cd
Amount of substance	mole	mol

* not required for the CCEA specification

Knowledge check 1

What is a base unit?

Knowledge check 2

State six SI base units.

SI derived units

Units for all other physical quantities are **derived units**, obtained from the base units by multiplying or dividing one base unit by one or more other units. For example:

$$\text{speed} = \frac{\text{distance}}{\text{time}} \qquad \text{units: } \frac{\text{metres}}{\text{seconds}}, \text{ m s}^{-1}$$

$$\text{charge} = \text{current} \times \text{time} \qquad \text{units: amperes} \times \text{seconds, A s}$$

Many derived units have their own special name. Those used in the specification are shown in Table 2.

Table 2 Derived units

Quantity	Derived unit (in terms of base units)	Name of derived unit	Symbol for unit
Frequency	s^{-1}	hertz	Hz
Force	$kg\,m\,s^{-2}$	newton	N
Pressure	$kg\,m^{-1}\,s^{-2}$	pascal	Pa
Energy	$kg\,m^2\,s^{-2}$	joule	J
Power	$kg\,m^2\,s^{-3}$	watt	W
Charge	A s	coulomb	C
Electric potential difference	$kg\,m^2\,s^{-3}\,A^{-1}$	volt	V
Resistance	$kg\,m^2\,s^{-3}\,A^{-2}$	ohm	Ω

Quantities that do not have a named derived unit are expressed in terms of the base unit combination. For example, velocity is measured in metres per second.

Examiner tip

The special names given to many derived units are the names of famous scientists. However, when used for the SI unit a lower case first letter must be used, for example hertz. The symbols in these cases do use a capital first letter, for example Hz.

You should never add an 's' to denote the plural, for example 5 Hz, *not* 5 Hzs.

Prefixes for SI units

Sub-multiples and multiples of both base and derived units exist. This is to cater for the very large and the very small values that exist with these quantities. There are approved prefixes and symbols (Table 3).

Table 3 Prefixes

Sub-multiple	Prefix	Symbol	Multiple	Prefix	Symbol
10^{-2}	centi	c	10^3	kilo	k
10^{-3}	milli	m	10^6	mega	M
10^{-6}	micro	μ	10^9	giga	G
10^{-9}	nano	n	10^{12}	tera	T
10^{-12}	pico	p			
10^{-15}	femto	f			

The symbol for kilo is an anomaly as, unlike other multiples, it is not a capital; this is because K is used for Kelvin.

Examiner tip

When writing a derived unit a space is left between the units for each quantity. For example, this distinguishes $m\,s^{-1}$ (meters per second) from ms^{-1}, which represents the inverse of milliseconds.

Knowledge check 3

Express 24 nm in standard form.

Examiner tip

When recording a large or small answer it is best to present it in standard form, that is, as a number between 1 and 10, times a power.

Equations and units

One of the conditions for an equation to be valid is that each of the terms in the equation must have the same base units. The equation is then said to be **homogeneous**. For example in:

$$v = u + at$$

both v and u are velocity with base units $m\,s^{-1}$. The product at has base units of $m\,s^{-2} \times s$, which also gives $m\,s^{-1}$. Therefore the equation is homogeneous or balanced.

This technique can be used to find the units of an unknown quantity in an equation and hence give a clue to the quantity.

Examiner tip

Homogeneity shows that an equation could be correct, but does not prove that it is correct.

Worked example

The unit of stress is the pascal. Express the pascal in base units.

Answer

The pascal, Pa, is the SI unit of pressure.

Recall equations for the quantity for which the base units are known:

$$\text{pressure} = \frac{\text{force}}{\text{area}}$$

$$\text{force} = \text{mass} \times \text{acceleration}$$

Identify the base units for these quantities:

mass: kg

area: m^2

acceleration: $m\,s^{-2}$

Combine these units to find the base units for pressure:

$$\text{base units of pressure} = \frac{\text{kg} \times \text{m}\,\text{s}^{-2}}{\text{m}^2} = \text{kg}\,\text{m}^{-1}\,\text{s}^{-2}$$

Examiner tip

You should know how to multiply and divide powers. To multiply, add the powers:
$k^x \times k^y = k^{(x+y)}$
To divide, subtract the powers:
$k^x \div k^y = k^{(x-y)}$

Worked example

In the equation for Newton's law of gravitation:

$$F = \frac{GMm}{r^2}$$

Find the base units of G.

Answer

base units of force $F = \text{kg}\,\text{m}\,\text{s}^{-2}$

base unit of masses M and m = kg

base unit of distance r = m

Change the subject of the equation to G:

$$\frac{Fr^2}{Mm}$$

Substitute in the base units for each term:

$$\text{base units of } G = \frac{(\text{kg}\,\text{m}\,\text{s}^{-2})(\text{m}^2)}{(\text{kg})(\text{kg})}$$

Simplify by combining and cancelling as you would with numbers:

$$\text{base units of } G = \text{kg}^{-1}\,\text{m}^3\,\text{s}^{-2}$$

Summary

- Physical quantities consist of a numerical magnitude and a unit.
- It is the units that distinguish one quantity from another.
- SI refers to the International System of Units. It is a coherent system based on seven base units.
- The seven base units (and their associated quantities) are: metre (length), kilogram (mass), second (time), ampere (electric current), kelvin (temperature), mole (amount of substance) and candela (luminous intensity).
- Other physical quantities required can be measured in derived units; these are formed by combining base units.
- For an equation to be balanced, the terms on each side of the equation must have the same base units.

Scalars and vectors

Physical quantities can be classified as either **scalar** or **vector** (Table 4).

Scalar quantities can be fully described by their magnitude and a unit. The direction has no significance — for example, body temperature is fully described as 37°C and the mass of a person as 75 kg.

Vector quantities are *not* fully described by *only* their magnitude and a unit — the situation is direction dependent. A force of 10 N acting on an object will produce a different outcome if it acts to the right compared with if it acted to the left. A vector quantity is only fully described when the direction, magnitude and unit are stated.

Table 4 Scalars and vectors

Scalars	Vectors
Mass	Weight
Temperature	Force
Energy	Momentum
Power	Acceleration
Volume	Electric current
Time	Moment of a force
Distance	Displacement
Speed	Velocity

Adding scalars and vectors

The addition of scalar quantities is non-problematic, it is a simple arithmetic sum. For example, the total mass of 2 kg and 3 kg is 5 kg. The increase in temperature from −5°C to 12°C is 17°C.

However, when two vectors are added, we need to take account of their direction as well as their magnitude. There are two methods — by scale drawing or, when they are perpendicular to one another, by use of Pythagoras' theorem.

The sum of two vectors is known as the **resultant**.

Examiner tip

Temperature and charge are scalar quantities, yet can have negative values. In these instances the negative denotes differences in size and type respectively, and not direction.

Knowledge check 4

State whether the following quantities are scalar or vector: work, density, charge, momentum.

Knowledge check 5

For a falling body, state three scalar and three vector quantities associated with its motion.

Examiner tip

Subtraction should be considered as the addition of the negative of the second vector. The negative vector is obtained by reversing the direction of the original vector: $\mathbf{z} = \mathbf{y} - \mathbf{x}$ or $\mathbf{z} = \mathbf{y} + (-\mathbf{x})$

Examiner tip

Giving the value of θ alone is not sufficient; you must give a compass bearing. For example, in Figure 1 these would be (a) east of south or (b) west of south.

Knowledge check 6

Two forces of magnitude 12 N and 5 N act on a body. What is:

(a) the minimum magnitude of the resultant force?

(b) the maximum magnitude of the resultant force?

(c) the value of the resultant force when they act at 90° to each other?

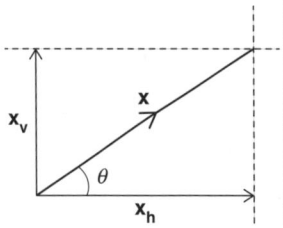

Figure 2

Knowledge check 7

Which is more effective, pushing or pulling a garden roller? Explain your answer.

Scale drawing

The length of the line, drawn to scale, represents the magnitude and it can be drawn in the direction of the vector. The start of the second vector is drawn from the tip of the first. This is known as the 'nose-to-tail' method.

So to add, or subtract, vectors, you transpose (move in space) the second vector so that you effectively have the action of vector **y** followed by the action of vector **x**. The resultant vector **z** and its direction can be found by direct measurement from the diagram (Figure 1).

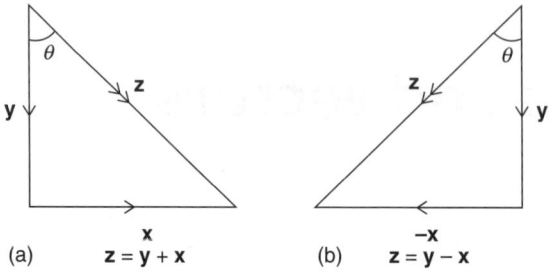

Figure 1

Pythagoras' theorem

To find the magnitude of the resultant use the equation:

$$\mathbf{x}^2 + \mathbf{y}^2 = \mathbf{z}^2$$

To find the direction of the resultant vector use trigonometry:

$$\theta = \tan^{-1}\left(\frac{\mathbf{x}}{\mathbf{y}}\right)$$

Resolution of vectors

Just as we can combine the action of individual vectors to find the single resultant vector, it is sometimes useful to split a vector into its perpendicular components.

Again, the horizontal and vertical components can be found by two methods — scale drawing and trigonometry (Figure 2).

Scale drawing

- The length of the line, drawn to scale, represents the magnitude and it can be drawn in the direction of the vector.
- Lines are drawn horizontally and vertically from the end of the vector.
- Lines are drawn horizontally and vertically from the start of the vector and stop when they meet the previous drawn lines. These lines are the horizontal and vertical components of the vector.

Trigonometry

$$\sin\theta = \frac{\mathbf{x}_v}{\mathbf{x}} \text{ and } \cos\theta = \frac{\mathbf{x}_h}{\mathbf{x}}$$

$$\mathbf{x}_v = \mathbf{x}\sin\theta \text{ and } \mathbf{x}_h = \mathbf{x}\cos\theta$$

Worked example

A man can row at a speed of $3\,m\,s^{-1}$. He wants to row directly across a river. The river current is $2\,m\,s^{-1}$. Find the direction in which the man must row and his resultant speed.

Answer

Construct a vector diagram (Figure 3).

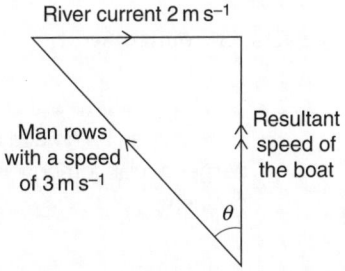

Figure 3

The man must row upstream to allow for the effect of the current.

$$\text{direction } \theta = \sin^{-1}\frac{2}{3}$$
$$= 42°$$

$$\text{resultant speed of the boat} = \sqrt{3^2 - 2^2}$$
$$= \sqrt{5}$$
$$= 2.2\,m\,s^{-1}$$

Worked example

Two forces of 3 N and 4 N act on a body B as shown in Figure 4.

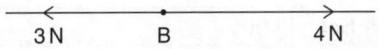

Figure 4

(a) State the magnitude and direction of the resultant force on B.

(b) The 4 N force is now turned through 90°, as shown in Figure 5. Find, by calculation, the new resultant force on body B. State the direction of this force relative to the horizontal.

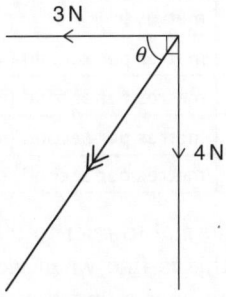

Figure 5

Knowledge check 8

A tennis player hits a ball with an initial speed of $12.0\,m\,s^{-1}$ at an angle of 30° to the horizontal. What are the initial horizontal and vertical components of the speed of the ball.

Answer

(a) resultant force = 4 – 3 = 1 N to the right

(b) resultant force = $\sqrt{4^2 + 3^2} = 5\,\text{N}$

$$\tan\theta = \frac{4}{3}$$

$$\theta = 53°$$

The resultant force acts 53° south of west.

Summary

- Scalar quantities have magnitude only.
- Vector quantities have magnitude and a direction.
- In a scale drawing the length of an arrow represents the magnitude of the vector and its direction the direction of the vector.
- A vector can be divided or resolved into two perpendicular components to assist in solving problems.

Linear motion

We are all familiar with journeys of one form or another and the quantities involved in describing them:

- How far away is it?
- How long will it take?
- Can't we go any faster?
- Are we there yet?

The quantities involved in describing motion are listed in Table 5.

Table 5

Quantity	Symbol	Unit	Vector or scalar
Distance	s	metres (m)	Scalar
Initial speed	u	metres per second (m s^{-1})	Scalar
Final speed	v	metres per second (m s^{-1})	Scalar
Average speed	$\langle v \rangle$	metres per second (m s^{-1})	Scalar
Time	t	seconds (s)	Scalar
Displacement	s	metres (m)	Vector
Initial velocity	u	metres per second (m s^{-1})	Vector
Final velocity	v	metres per second (m s^{-1})	Vector
Average velocity	$\langle v \rangle$	metres per second (m s^{-1})	Vector
Acceleration	a	metres per second^2 (m s^{-2})	Vector

For everyday descriptions it is normal to refer to the scalar quantities and a loose interpretation of acceleration. This is fine when we are dealing with motion in a straight line, in one direction, but once a direction change occurs we must use the vector quantities.

Definitions of quantities

Speed is the rate of change of distance or the distance moved per second:

$$\text{speed} = \frac{\text{distance travelled}}{\text{time taken}}$$

During a journey the speed may change, hence the concept of average speed for the whole journey:

$$\text{average speed} = \frac{\text{total distance travelled}}{\text{total time taken}}$$

Velocity, a vector quantity is defined in terms of displacement:

$$\text{velocity} = \frac{\text{displacement}}{\text{time taken}}$$

$$\text{average velocity} = \frac{\text{total displacement}}{\text{total time taken}}$$

During a journey when the velocity is changing, either in magnitude (speeding up or slowing down) or in direction only, it is accelerating:

$$\text{acceleration} = \frac{\text{change in velocity}}{\text{time taken for the change}}$$

The need for the vector and scalar distinction is best illustrated with a simple example.

Examiner tip

When an object is slowing down, acceleration will have a negative value. This is known as deceleration or retardation.

Worked example

A car completes a half lap of a circular track of circumference 6000 m in 150 s (Figure 6). Assume that the car travels at a constant speed.

(a) What is the constant speed?

(b) What is the change in velocity after half a lap?

(c) What is the acceleration during the half lap?

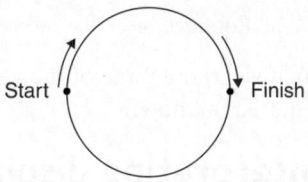

Figure 6

Answer

(a) $\text{speed} = \frac{\text{distance travelled}}{\text{time taken}}$

$\text{speed} = \frac{3000}{150} = 20\,\text{m s}^{-1}$

(b) Note that the directions the car is travelling in at the start and finish are *opposite*. We use plus and minus signs to denote this difference.

initial velocity, $u = +20\,\text{m s}^{-1}$

final velocity, $v = -20\,\text{m s}^{-1}$

change in velocity $= v - u = (-20) - (20) = -40\,\text{m s}^{-1}$

(c) $\text{acceleration} = \frac{\text{change in velocity}}{\text{time taken}}$

$\text{acceleration} = \frac{40}{150} = 0.27\,\text{m s}^{-2}$

Examiner tip

At every point on the track the speed is the same; however, the velocity is continually changing, because the direction is changing.

Examiner tip

Which direction you choose as + is up to you, but be consistent throughout your answer.

Knowledge check 9

A car starting from rest accelerates in a straight line at $4\,m\,s^{-2}$ for 5 s. How far has the car travelled in this time?

Knowledge check 10

A car goes from rest to 60 mph in 4.5 s. Calculate its acceleration.

(1 mile = 1.6 km)

Knowledge check 11

A body travelling at $10\,m\,s^{-1}$ decelerates at $4\,m\,s^{-2}$. How far does it travel before coming to a stop?

Knowledge check 12

A car accelerates from $5\,m\,s^{-1}$ to $25\,m\,s^{-1}$ in 8 s. How far does it travel in this time and what is the value of its acceleration?

Examiner tip

It is possible to draw distance–time and speed–time graphs, but their use is limited as they cannot convey direction reversal. (Displacement–time and velocity–time graphs *can* convey direction reversal.)

So, in this example, even though there has been no change of speed there has been a small acceleration. Refer to circular motion.

Linear motion is motion in a straight line, so the only direction change we will deal with will be up and down or right and left. We can use + and – signs to indicate opposite directions.

Equations of motion

These equations are limited to motion of an object with **constant acceleration in a straight line**. They are based on the basic definitions already given above and one additional point.

For an object moving with constant acceleration the average velocity will be halfway between the initial and the final velocities.

$$\langle v \rangle = \tfrac{1}{2}(u + v)$$

The four equations of motion are sometimes referred to as the 'suvat' equations, because of the terms used:

Equation 1:	$v = u + at$	(no s involved)
Equation 2:	$s = \tfrac{1}{2}(u + v)t$	(no a involved)
Equation 3:	$s = ut + \tfrac{1}{2}at^2$	(no v involved)
Equation 4:	$v^2 = u^2 + 2as$	(no t involved)

So, if we know three of the variables, by using the correct equation we can find the required unknown.

Interpreting displacement–time and velocity–time graphs

A useful way of representing a journey is with a displacement–time or velocity–time graph.

As well as effectively giving an overall summary of the nature of the journey we can also calculate the unknown values from the appropriate graph.

Displacement–time graph

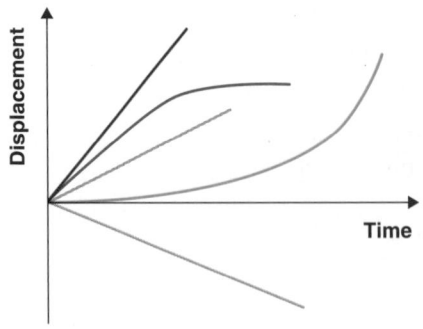

Figure 7

In Figure 7 the coloured lines represent different motions:

- grey — object moving at a constant velocity
- blue — object moving at a greater constant velocity
- orange — object moving at a constant velocity but in the opposite direction
- green — object starting from rest and speeding up
- red — object has a starting velocity and is slowing down to a stop

All this information is based on one principle:

The gradient or slope of a displacement–time graph is equal to the velocity of the object.

It is therefore possible to calculate a value for the velocity from the displacement–time graph.

Examiner tip

Any journey will be a combination of the lines in Figure 7.

Velocity–time graphs

The same descriptions of motion from above can be represented on a velocity–time graph (Figure 8).

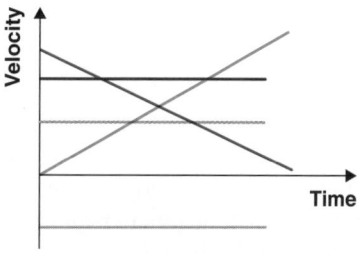

Figure 8

Here the principle is as follows:

The gradient or slope of a velocity–time graph is equal to the acceleration of the object.

In addition to being able to determine the value of the acceleration from a velocity–time graph, we can evaluate the distance travelled or the total displacement.

Consider the example of a ball thrown vertically upwards, from waist high, and allowed to fall to the ground. The velocity–time graph for the ball is shown in Figure 9. Note the sign convention of up as +.

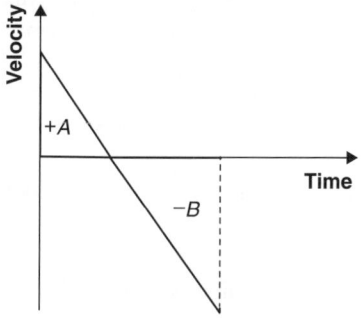

Figure 9

The area between the velocity–time plot and the time axis is equal to the distance travelled by the ball.

distance travelled = $A + B$

If we consider the sign associated with the value of the area then we get the displacement:

displacement of the ball = *A* – *B*

Note that in our example area B > area A; the difference is the value from the waist to the ground.

Another possible shape for the velocity–time graph is a curve or non-linear shape. This represents changing acceleration. A classic example of non-uniform acceleration occurs in a parachute jump. The change in the acceleration is a consequence of the change in the force acting on the body (Figure 10).

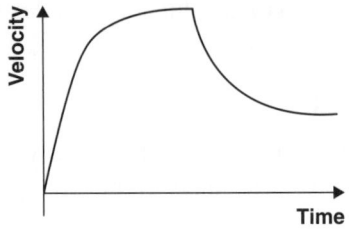

Figure 10

Initially the parachutist is in free-fall but as his velocity increases the upward force of air-resistance increases, causing the acceleration to decrease (slope decreasing). When no resultant force acts the acceleration is zero (horizontal line). The parachute is opened, increasing the frictional force, and the parachutist decelerates (negative slope) until once again the resistive force is equal and opposite to the weight and he falls at a constant 'terminal' velocity (horizontal line).

Examiner tip

The specification does not demand a specific method so you need to know, in detail, the method you use in school, and also the general principles required to measure acceleration.

Experiment to measure the acceleration of free fall

Because of the variable force of attraction of the Earth on objects of different mass they will all fall with the same uniform acceleration. This acceleration is called the **acceleration of free fall** and is represented by the symbol g.

There are a variety of methods used to determine g.

Use of $s = ut + \frac{1}{2}at^2$

If we ensure the object falls from rest, i.e. $u = 0$, the equation will reduce to:

$$s = \tfrac{1}{2}gt^2$$

In an experiment we time the fall of an object, from rest, through a range of different distances, noting the time of fall in each case. A graph of s against t^2 will be linear, with a slope equal to a value of $\frac{1}{2}g$.

As the time values are so small it is usual to use an electrical timer, which is triggered to commence timing as a steel ball is released, and stop timing as the ball breaks a circuit by opening a trap door (Figure 11).

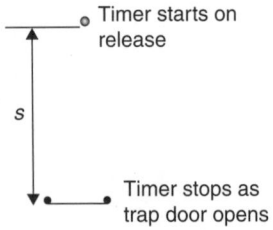

Figure 11

Use of $a = \frac{v - u}{t}$

Here we use a card of known length to fall through two light gates attached to a datalogger and computer. The software will measure time values to one-hundredth of a second. Three times are automatically recorded:

- the time for the card to fall through gate A, t_A
- the time for the card to fall through gate B, t_B
- the time for the card to fall from gate A to gate B, t_{AB}

The length, s, of the card is measured and entered into the computer.

initial velocity, $u = \frac{s}{t_A}$

final velocity, $v = \frac{s}{t_B}$

acceleration of free fall, $g = \frac{(v - u)}{t_{AB}}$

With the datalogger and light gate system it is possible to obtain numerous results very quickly, without the need for graphical analysis. This allows the opportunity for numerous repeats and variation of factors — for example, changing the mass of the card.

Worked example

A jet aircraft accelerates from rest for 30.0 seconds, leaving the ground at a speed of 285 kilometres per hour.

(a) Find the take-off speed in metres per second.

(b) Calculate the average acceleration of the aircraft during take-off.

(c) Calculate the length of runway used by the aircraft during take-off.

Answer

(a) $285\,\text{km h}^{-1} = 285 \times 1000\,\text{m h}^{-1}$

$$= \frac{285\,000}{60 \times 60\,\text{m s}^{-1}}$$

$$= 79.2\,\text{m s}^{-1}$$

(b) acceleration, $a = \frac{v - u}{t}$

$$a = \frac{79.2 - 0}{30} = 2.64\,\text{m s}^{-2}$$

(c) displacement, $s = \frac{1}{2}(u + v)t$

$$s = \frac{1}{2}(0 + 79.2) \times 30 = 1188\,\text{m}$$

Examiner tip

When possible, take multiple readings and quote the mean. The uncertainty in your result is the difference between the mean and the extreme reading.

Knowledge check 13

A ball is thrown vertically upward from the ground with a velocity of $28\,\text{m s}^{-1}$. After how many seconds will it strike the ground?

Worked example

A cricketer throws a ball vertically upwards with a velocity of $14.0\,\mathrm{m\,s^{-1}}$ and catches it as it falls.

(a) How high does the ball go?

(b) How long does it take to reach this height?

(c) How long is the ball in the air?

Answer

(a) Consider the flight of the ball up to the highest point.
At the highest point the ball is momentarily at rest, therefore:

$v = 0$

initial velocity, $u = 14.0\,\mathrm{m\,s^{-1}}$, upwards

The acceleration is constant throughout and in the opposite direction from the initial velocity, therefore:

$a = g = -9.81\,\mathrm{m\,s^{-2}}$

Use the equation:

$v^2 = u^2 + 2as$

Rearranged:

$$s = \frac{v^2 - u^2}{2a}$$

$$s = \frac{0 - 14.0^2}{2(-9.81)}$$

$$s = \frac{-196}{-19.62} = +9.99\,\mathrm{m}$$

(b) Use the equation:

$v = u + at$

Rearranged:

$$t = \frac{v - u}{a}$$

$$t = \frac{0 - 14.0}{-9.81} = 1.43\,\mathrm{s}$$

(c) As the ball is caught at the same point from which it was propelled the overall displacement is zero. This allows direct solution using the equation:

$s = ut + \frac{1}{2}at^2$

$0 = 14.0t + \frac{1}{2}(-9.81)t^2$

Rearranging:

$$14.0t = 4.9t^2 \text{ or } t = \frac{14.0}{4.9} = 2.86\,\mathrm{s}$$

Examiner tip

Note that the answer to part (c) is twice that of the answer to part (b). This is no surprise as the motion has symmetry. But this will only be true when the net vertical displacement is zero.

Worked example

A ball is dropped from rest onto the ground and is allowed to bounce before being caught at the highest point. The velocity–time graph for the motion of the ball is shown in Figure 12.

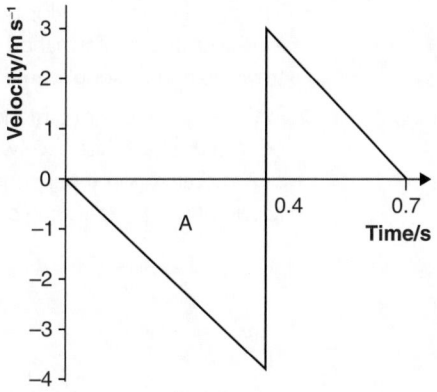

Figure 12

(a) (i) With what velocity does the ball hit the ground?
(ii) With what velocity does it leave the ground?

(b) Use the graph to find the acceleration of free fall.

(c) From what height was the ball released?

Answer

(a) At 0.4 s the velocity of the ball suddenly changes from a negative to a positive value. This indicates a change of direction, which occurs when the ball hits the ground.
(i) The velocity as it hits the ground is $-4.0\,\mathrm{m\,s^{-1}}$ (read directly from the graph).
(ii) The velocity as it leaves the ground is $+3.0\,\mathrm{m\,s^{-1}}$ (read directly from the graph).

(b) The acceleration is equal to the slope of the velocity–time graph.
Two slopes can be determined from the graph.
As it is falling:

$$a = g = \frac{\Delta v}{t} = \frac{-4 - 0}{0.4} = -10\,\mathrm{m\,s^{-2}}$$

As it is rising:

$$a = g = \frac{\Delta v}{t} = \frac{0 - 3.0}{0.3} = -10\,\mathrm{m\,s^{-2}}$$

average value of acceleration, $g = -10\,\mathrm{m\,s^{-2}}$
(The negative indicates a downward direction.)

(c) The distance travelled is equal to the area between the velocity–time graph and the time axis.
We want the distance travelled from when the ball is released and when it hit the ground, i.e. $t = 0$ to $t = 0.4$ s.
We want the area of triangle A, labelled in the diagram.

area of triangle = ½ × base × perpendicular height

area of triangle A = ½ × 0.4 × 4.0 = 0.8 m

Examiner tip
When the ball is falling it has negative velocity. When the ball is rising it has positive velocity.

Summary

- Displacement, a vector quantity, is defined as the change in position from a starting or reference point.
- Speed is defined as the distance moved per second.
- Velocity is defined as the rate of change of displacement.
- Acceleration is defined as the rate of change of velocity.
- For linear motion problems we can use + and − to indicate opposite directions.
- For bodies moving with constant acceleration in a straight line:

  $v = u + at$

  $s = \frac{1}{2}(u + v)t$

  $s = ut + \frac{1}{2}at^2$

  $v^2 = u^2 + 2as$
- The gradient of a displacement–time graph is equal to the velocity of a body.
- The gradient of a velocity–time graph is equal to the acceleration of a body.
- The area between a velocity–time plot and the time axis gives us the information to determine distance travelled and the displacement of the body.
- In the absence of air resistance, a body falling freely near to the Earth's surface will experience the same acceleration, independent of its mass. The acceleration, g, has a value of $9.81\ m\,s^{-2}$.

Projectiles

Description of projectile motion

Projectile motion is motion in two dimensions; the object is acted on by a constant force in one dimension and no force in the other dimension.

So far we have considered the motion of objects in a straight line, but another common situation exists — **projectile motion**.

Any object propelled at an angle to the vertical near the Earth's surface is an example of projectile motion.

A charged particle moving at an angle to a uniform electric field is another example of projectile motion.

In projectile motion the particles will follow a curved path specifically known as a **parabola**.

Explanation of projectile motion

Examiner tip

The vertical force has no component in the horizontal plane and cannot therefore cause any change to its horizontal velocity component.

Consider the most common case of a thrown particle. It is subject to a vertical gravitational force — its weight. The horizontal force of air resistance is neglected.

Whilst moving forward, the particle also moves downwards under the effect of gravity. The resultant path of the projectile is in effect a combination of two motions — horizontal and vertical.

This allows us to work with the horizontal and vertical aspects of motion separately and, as they are in straight lines, apply the equations of motion.

Note that the time of flight is the same for both.

To find the actual velocity of the particle we simply add the components vectorially (Figure 13).

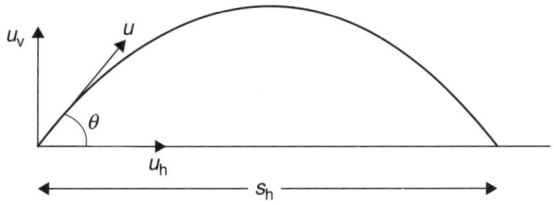

Figure 13

The terms in Table 6 can be applied to all projectile questions.

Table 6

Horizontal component	Vertical component
$u_h = u\cos\theta$	$u_v = u\sin\theta$
$a_h = 0$	$a_v = g$
s_h = range	$s_v = \Delta h$ = change in height
t_h = time of flight	t_v = time of flight

Let us apply this technique to a simple problem.

Worked example

A stone is thrown horizontally from the top of a cliff 60 m high, with a velocity of $12\,\mathrm{m\,s^{-1}}$.

(a) How long does it take to reach the ground below?

(b) How far from the base of the cliff does it hit the ground?

(c) What is its speed and direction on hitting the ground?

Answer

Apply this to the general horizontal and vertical terms. Note that for this question we have taken the down direction as positive.

Horizontal component	Vertical component
$u_h = u\cos\theta = 12\cos 0 = 12\,\mathrm{m\,s^{-1}}$	$u_v = u\sin\theta = 0$
$a_h = 0$	$a_v = g = 9.81\,\mathrm{m\,s^{-2}}$
s_h = range?	$s_v = \Delta h$ = change in height = 60 m
t_h = time of flight	t_v = time of flight

(a) Use the vertical component values and $s = ut + \frac{1}{2}at^2$:

$$60 = 0 + \tfrac{1}{2}(9.81)(t^2)$$
$$t^2 = 12.2$$
$$t = 3.5\,\mathrm{s}$$

(b) Use the horizontal component values and $s = \frac{1}{2}(u + v)t$. The horizontal component of the velocity does not change, so $u = v$:

$$s_h = \tfrac{1}{2}(12 + 12) \times (3.5) = 42.0\,\text{m}$$

(c) Acceleration due to gravity means that the vertical component of velocity is increasing. Use the equation $v = u + at$:

$$v_v = u_h + gt$$

$$v_v = 0 + (9.81 \times 3.5) = 34.3\,\text{m s}^{-1}$$

The resultant or actual velocity of the stone is the vector sum of the two components:

$$v_r = \sqrt{v_h^2 + v_v^2} = \sqrt{12.0^2 + 34.3^2} = 36.4\,\text{m s}^{-1}$$

The direction of the stone can be found by reference to the vector diagram (Figure 14).

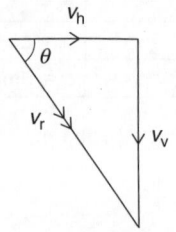

Figure 14

$$\tan\theta = \frac{v_v}{v_h} = \frac{34.3}{12.0}$$

$\theta = 70.7°$ below the horizontal

Knowledge check 14

What is the name given to the flight path followed by a projectile?

Knowledge check 15

An object is projected with a velocity of $10.0\,\text{m s}^{-1}$ at an angle of 30° to the horizontal. State the values of the acceleration, vertical component of the velocity and horizontal component of the velocity (a) at the start and (b) at the top of its flight.

Knowledge check 16

A projectile lands at the same vertical height from which it is launched. It is in the air for 6.8 s. How long did it take to reach its highest point?

Summary

- Projectiles follow a parabolic path.
- A projectile has a uniform velocity in one direction and a uniform acceleration in a perpendicular direction.
- The equations of motion can be applied independently to the two perpendicular components of the projectile motion.

Newton's laws of motion

We have described the motion of objects by graphical means and through the equations of motion. But now we consider what causes objects to move.

- From our everyday experiences it is obvious that force is required.
- A body requires a force to move it from rest or bring it to rest.
- A body requires a force to change its speed or direction.

It was Isaac Newton who cleared up the misconception that a force was needed to keep a body moving with a constant velocity. He stated that no **resultant** force acted; rather, a force was needed to overcome unseen frictional, resistive forces.

Examiner tip

Forces are vector quantities; the SI unit of force is the newton (N).

Newton summarised his theory of motion in three laws. (There are several versions of the wording for these laws.)

1 **A body will remain at rest or continue to move with a constant velocity unless acted on by a resultant force.**
2 **For a body of constant mass, its acceleration is directly proportional to the resultant force applied and in the direction of the resultant force.**
3 **Whenever one body exerts a force on another, the second body exerts an equal and opposite force on the first body.**

Newton's laws are more easily visualised when we consider a near friction-free environment, for example on an ice rink.

1 Once moving, a skater continues to glide with the same speed in a straight line.
2 When a force is applied to the skater their velocity changes; acceleration takes place.
3 If two stationary skaters, A and B, face one another and skater A pushes skater B, both are seen to move in opposite directions. To the action of A on B there is a reaction of B on A.

Examiner tip
The first and second law both comply with the equation:
$F_{resultant} = ma$

Examiner tip
It is important to emphasise in the third law that the acting force and the reacting force act on different bodies.

Friction: a force that opposes motion

Frictional forces are the unseen forces that need to be considered if everyday situations are to be seen to obey Newton's laws. These frictional forces:

- will always oppose motion
- are reactionary forces — they will only act to oppose action forces
- give rise to the development of heat energy
- often occur when a moving body is in contact with another medium
- can be helpful, for example the grip needed when walking and braking
- can be a hindrance, for example the wear in moving engine parts

Examiner tip
Use of a lubricant indicates that the friction effect is not helpful.

Worked example

A body of weight 30.0 N is suspended from a fixed point by a vertical string. Another force pulls the body sideways along the horizontal until the string is at an angle of 35° to the vertical.

(a) Sketch a diagram to show all the forces acting on the object. Clearly label and name the forces and indicate their direction.

(b) Calculate the magnitude of the tension in the string.

(c) Find the value of the horizontal pulling force.

Examiner tip
Read all the parts of the question first — often clues appear later in the question, in this case the name of one of the forces. Also, the overall focus of the question is more readily understood.

Answer

(a)

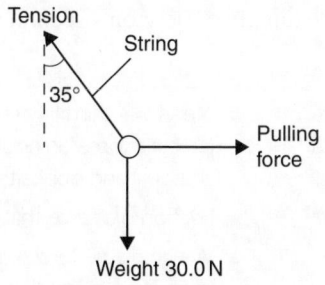

Figure 15

Examiner tip
When dealing with multiple forces acting on a body at different angles, resolve them into two components perpendicular to one another.

(b) The body is at rest, so the sum of the components in the horizontal and vertical planes must be zero.
Consider the vertical components:

$$F_{resultant} = 0 = T\cos 35° - W$$

Therefore:

$$T\cos 35° = 30.0 \text{ and } T = 36.6\,\text{N}$$

(c) Consider the horizontal components:

$$F_{resultant} = 0 = T\sin 35° - F_{pull}$$

Therefore:

$$F_{pull} = (36.6\sin 35) = 21.0\,\text{N}$$

Knowledge check 17

The total mass of a lift and its passengers is 1000 kg. The lift ascends at a constant speed of $1\,\text{m s}^{-1}$. What is the tension in the cable pulling the lift?

Knowledge check 18

A car of mass 1200 kg accelerates uniformly from rest to a velocity of $12\,\text{m s}^{-1}$ in 24 s. Calculate the force that produced this acceleration.

Knowledge check 19

The tension in the towrope pulling a water skier of mass 80 kg is 560 N.

What is the value of the frictional force acting on the skier if his acceleration is $1.5\,\text{m s}^{-2}$?

Worked example

A pushing force of 5.0 N is applied to a body of mass 2.5 kg.

The body accelerates at $1.8\,\text{m s}^{-2}$.

Find the frictional force that is acting on the body.

Answer

$$F_{resultant} = F_{push} - F_{friction} = ma$$

$$F_{resultant} = 5.0 - F_{friction} = (2.5 \times 1.8)$$

$$F_{friction} = 5.0 - 4.5 = 0.5\,\text{N}$$

Weight and Newton's second law

Any body close to the surface of the Earth will fall to Earth with the same acceleration, g, known as the acceleration due to gravity or the acceleration of free fall.

It was Galileo who first confirmed that g was independent of mass. For an object of larger mass to fall with the same acceleration as a lighter mass, a greater force must act.

According to Newton's second law:

$$F = mg$$

This force with which an object is pulled towards the Earth is called its **weight**, ***W***. So $W = mg$ is one specific example of Newton's second law.

Summary

- Newton's first law states: a body will remain at rest or continue to move with a constant velocity unless acted on by a resultant force.
- Newton's second law states: for a body of constant mass, its acceleration is directly proportional to the resultant force applied and in the direction of the resultant force.
- Newton's third law states: whenever one body exerts a force on another, the second body exerts an equal and opposite force on the first body.
- Friction is a force that opposes motion.
- Weight is the force with which a body is pulled towards the Earth's surface and is given by $W = mg$. This is an application of Newton's second law.

Principle of moments

Rather than cause a body to move in accordance with Newton's law, a different outcome will occur if the object is fixed at a point. The force will cause the object to turn about the point.

Definition of the moment of a force

A familiar example of this concept is seen in the use of a spanner to remove a wheel nut. You apply a force at the end of the spanner to make the nut rotate on the bolt.

By extending the length of the spanner you can reduce the force needed to release the nut. It is clear that there are two factors that determine the turning effect:

- the distance from the pivot
- the size of the force

The turning effect of a force is called the **moment** of the force. The SI unit of the moment of a force is the newton metre, N m.

$$\text{moment of force} = F \times d$$

The **moment** of a force about a point is defined as the product of the force and the perpendicular distance of the line of action of the force from the point.

Worked example

A mechanic has to apply a force of 150 N to just loosen the wheel nut on a car. He uses the garage spanner, which is 50 cm long.

What force would the owner of the car have to apply in the same situation if he used his own spanner, which is 20 cm long?

Answer

Calculate the moment required:

$$\text{moment of force} = F \times d = 150 \times 0.5 = 75\,\text{N m}$$

The same moment is required:

$$\text{moment of force} = 75 = F \times 0.2$$

$$\text{force applied using short spanner} = F = \frac{75}{0.2} = 375\,\text{N}$$

Examiner tip

The moment is a vector quantity having directional sense — referred to as clockwise or anticlockwise.

The principle of moments

When a body is acted on by more than one force and is in rotational equilibrium, then the turning effect of the forces must have cancelled each other. The most familiar example of this concept is the see-saw.

A long plank is fixed or pivoted about a centre point and two children sit at either end. The weight of the children provides the force.

From our childhood experiences we can again confirm that the two factors that determine the size of the turning effect are:

- the distance from the pivot
- the size of the force

The **principle of moments** states that when an object is in rotational equilibrium, the sum of the clockwise moments about a point is equal to the sum of the anticlockwise moments about the same point.

In addition we can intuitively calculate the value of the unknown quantity to achieve a balance. This is based on the **principle of moments**.

Worked example

Some weights are hung from a rod of negligible mass, as shown in Figure 16. The rod is pivoted as shown. Find the value of the force that must be applied at the end shown to balance the rod.

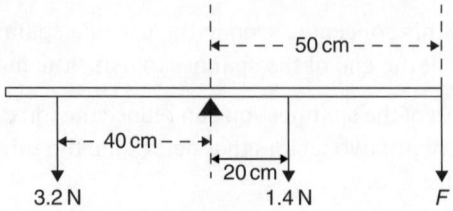

Figure 16

Answer

sum of clockwise moments = $(1.4 \times 0.20) + (0.50 \times F)$

sum of anticlockwise moments = (3.2×0.40)

Applying the principle of moments:

$$(1.4 \times 0.20) + (0.50 \times F) = (3.2 \times 0.40)$$

Rearranged:

$$0.50F = (3.2 \times 0.40) - (1.4 \times 0.20)$$

Therefore:

$$F = 2.0\,\text{N}$$

Centre of mass/gravity

In the worked example above we used a light rod. This was so that in the question we could ignore the mass and subsequently the weight of the rod.

The weight of a body when pivoted about a point will produce an unbalanced moment, except when the point is at the centre of mass/gravity. When pivoted about this point all the contributions to weight of the body produce moments that cancel; the clockwise and anticlockwise moments are equal. In mathematical terms, it is as if all the mass/weight of the object can be treated as acting through the centre of mass/gravity.

For a uniform body the centre of mass/gravity will be at the centre of the body. For a non-uniform body it will be closer to the heavy end, as there is a greater mass contribution from this end.

So in questions on moments we depict the weight of the body as a single force acting vertically down through the centre of gravity.

Worked example

To find the weight of a uniform metre rule a student set up the arrangement shown in Figure 17. The metre rule was pivoted at the 30 cm mark and a weight of 2.2 N was moved along its length until the ruler was balanced. The weight was at the 19 cm mark when the ruler was balanced.

Use the information given to calculate the weight of the metre rule.

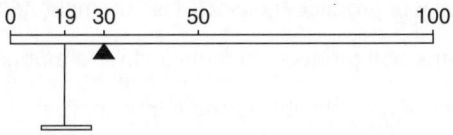

Figure 17

Answer

The body is in rotational equilibrium so the principle of moments applies:

sum of clockwise moments = sum of anticlockwise moments

The weight of the ruler produces the clockwise moment, and because the body is uniform this can be treated as acting at the 50 cm mark.

clockwise moments = $W \times (50 - 30) = 20\,W$

anticlockwise moments = $2.2 \times (30 - 19) = 24.2$

Therefore:

$20\,W = 24.2$

$W = 1.21$ N

The weight of the ruler is 1.2 N.

Worked example

A uniform plank AB, 16 m long and weighing 240 N, rests on two supports as shown in Figure 18. The support at X is fixed while the support at Y can be moved. A boy weighing 360 N starts at point X and walks towards B. How far from the right end (B) of the plank should support Y be placed so that he can just reach this end of the plank without causing it to tip?

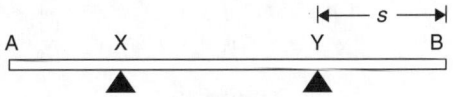

Figure 18

Answer

When the plank is just about to tip there will be no contact at support X and Y will be the only pivot.

Let Y be a distance s from end B.

Take moments about Y.

The weight of the boy will produce the clockwise moment $360 \times s$.

The weight of the plank will produce the anticlockwise moment $240 \times (8 - s)$.

In rotational equilibrium the principle of moments applies. Therefore:

$$360 \times s = 240 \times (8 - s)$$

$$360s + 240s = 1920$$

$$600s = 1920$$

$$s = 3.2\,\text{m}$$

Vertical equilibrium

Where a body is pivoted or supported, and is in rotational equilibrium under the action of a number of forces, Newton's first law will apply, as well as the principle of moments.

The resultant force acting must equal zero; the sum of the upward forces equals the sum of the downward forces.

The upward forces are supplied by the reaction forces at the pivot or support points.

Worked example

A uniform beam, of weight 240 N and length 9.6 m, is supported on two pillars, X and Y, 6.0 m apart. The beam projects 2.0 m from pillar X, as shown in Figure 19.

Calculate the support force of each pillar on the beam.

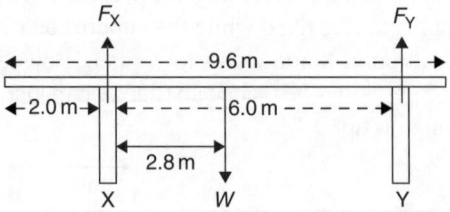

Figure 19

Examiner tip

In these calculations you can always eliminate the turning effect of an unknown force by taking moments about the point through which it acts.

Answer

The structure is in rotational equilibrium so the principle of moments applies.

Take moments about pillar X:

- F_X produces no turning effect about X.

- W, the beam weight, produces the clockwise moment.
- F_Y produces an anticlockwise moment.

clockwise moments = anticlockwise moments

$(240 \times 2.8) = (F_Y \times 6.0)$

Therefore:

$F_Y = 112\,\text{N}$

The structure is in vertical equilibrium, so $F_{\text{resultant}} = 0$.

forces up = forces down

$F_X + F_Y = W = 240\,\text{N}$

As $F_Y = 112\,\text{N}$:

$F_X = 240 - 112 = 128\,\text{N}$

Examiner tip

Note that $F_X > F_Y$; this is what we expect as the centre of mass is closer to pillar X. Always look at the answer and check that it makes sense in light of what you already know.

Knowledge check 20

A nut needs a moment of 4 N m to turn it. What force is required when using a spanner of length (a) 10 cm, (b) 20 cm?

Knowledge check 21

A crowbar 2.0 m long is levered about a point 0.20 m from its tip. What is the smallest force that must be applied to raise a load of 80 kg at the tip?

Summary

- The moment of a force about a point is defined as the product of the force and the perpendicular distance of the line of action of the force from the point.
- The principle of moments states that for a body in rotational equilibrium the sum of the clockwise moments about a point is equal to the sum of the anticlockwise moments about the same point.
- In addition, for complete equilibrium the resultant force in any direction must be zero.
- The centre of mass/gravity of a body is the point at which all the mass/weight can be treated as acting.

Work done, potential and kinetic energy

Work done

Work is done on a body when a force acting on the body makes it move. In doing so, energy is transferred to the body and takes a different form.

Doing work is a basic concept. Imagine stacking a shelf; it is clear to us that we are doing work and the amount of work needed for the task depends on:

- the height of the shelf
- the weight of the object being lifted and hence the force needed to lift it

This forms the basis of the work done equation:

work done = force × distance moved in the direction of the force

The unit of work is the **joule** (J).

1 **joule** of work is done when a force of 1 newton moves its point of application by 1 metre, in the direction of the force.

Worked example

Figure 20 shows a force of 60 N acting on a paving slab resting on a horizontal surface. Find, by calculation, the work done in moving the slab 6.0 m along the surface.

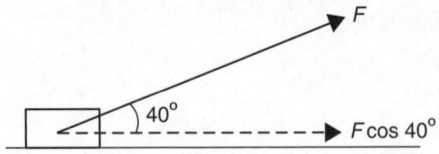

Figure 20

Answer

Find the component of the force acting in the direction of the movement:

work done = $F\cos\theta \times d = 60\cos 40° \times 6.0 = 276\,\text{J}$

Energy is the capacity to perform a task, whilst work is the 'doing' or carrying out of the task. The energy will change from one form into another.

Gravitational potential energy is the energy possessed by an object due to its raised position above the Earth's surface.

Energy transfer between gravitational potential energy and kinetic energy

In order to do work, or perform a task, we require an **energy** source.

Consider our simple shelf-stacking example; some chemical energy (derived from food) from the person stacking is converted into **gravitational potential energy**.

change in gravitational potential energy, $\Delta\text{GPE} = mg\Delta h$

If the object then fell from the shelf back to the floor the gravitational potential energy would be converted or transferred, progressively, into the energy of movement, known as **kinetic energy**.

change in kinetic energy, $\Delta\text{KE} = \frac{1}{2}mv^2 - \frac{1}{2}mu^2$

where u = initial velocity, in this case zero, and v = final velocity, in this case the velocity with which it hits the ground.

kinetic energy of an object (moving with a velocity v), $\text{KE} = \frac{1}{2}mv^2$

The unit of any form of energy will be the same as that of work done — the joule, J.

Examiner tip

As with any derived unit, when calculating the energy in joules we must use the SI numerical values for all quantities: mass in kg, velocity in m s^{-1}, g in m s^{-2} and Δh in m.

Conservation of energy

This simple example illustrates a fundamental principle of physics known as the **principle of the conservation of energy**.

The **principle of the conservation of energy** states that energy can neither be created nor destroyed, but can be changed in form.

Worked example

A slate of mass 2.0 kg falls from a roof 7.0 m above the ground. Assuming it falls from rest, calculate its kinetic energy as it hits the ground and its speed.

Answer

$\Delta KE = \Delta GPE$

$\Delta KE = mg\Delta h = 2.0 \times 9.81 \times 7.0 = 137.3\,J$

$\Delta KE = \frac{1}{2}mv^2 - \frac{1}{2}mu^2$

In this example $u = 0$.

$\Delta KE = \frac{1}{2}mv^2$

$137.3 = \frac{1}{2} \times 2.0 \times v^2$

Therefore:

$v^2 = 137.3$

$v = 11.7\,m\,s^{-1}$

Worked example

The car at the top of one slope on a roller coaster is travelling at $4.0\,m\,s^{-1}$, as shown in Figure 21. It continues down the slope and up to the top of the next one. Ignoring losses due to friction, calculate the speed of the car at the top of the second slope.

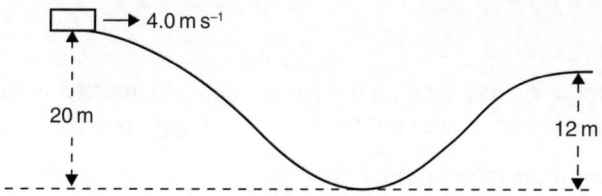

Figure 21

Answer

At the top of the first slope E_{total} = GPE and KE.

At the bottom of the first slope the GPE has been converted to an increased KE.

At the top of the second slope some of this KE has been converted back to GPE.

Overall there has been a net change of GPE to KE as the second slope is lower.

$\Delta GPE = \Delta KE$

$mg\Delta h = \frac{1}{2}mv^2 - \frac{1}{2}mu^2$

m cancels throughout:

$g\Delta h = \frac{1}{2}v^2 - \frac{1}{2}u^2$

$9.81 \times (20 - 12) = \frac{1}{2}(v^2 - 16)$

$v^2 = 173.0$

$v = 13\,m\,s^{-1}$

Power

Power is defined at the rate at which work is done or energy transferred from one form to another.

When work is done or a task completed we are often concerned with how quickly it has been carried out. The rate at which the work is done defines **power**:

$$\text{power} = \frac{\text{work done}}{\text{time taken}}$$

This also means that the more power available, the quicker a task can be carried out.

The SI unit of power is the watt, W. Like energy, it is a scalar quantity. This means, for example, that a 5 W motor will transfer energy at the rate of 5 J every second.

Examiner tip

The letter W can stand for weight, work done or the unit watts. Symbols for variable quantities such as weight or work done are expressed in *italics*, in this case *W*, while units are in normal text. So W (without italics) stands for watts. This difference is unseen when we record data, so be careful.

There is another useful form of the power equation, found by substituting in the equation for work done:

$$\text{power} = \frac{F \times d}{t}$$

But:

$$\frac{d}{t} = v$$

Therefore:

$$\text{power} = \text{force} \times \text{velocity}$$

$$P = Fv$$

Worked example

An electric motor is used to pull a boat up a ramp. The tension in the connecting cable is 600 N. The boat moves up the ramp at a steady speed of $0.6\,\text{m}\,\text{s}^{-1}$.

Find the power of the motor used.

Answer

$P = Fv$

$P = 600 \times 0.6 = 360\,\text{W}$

Efficiency

Efficiency is the ratio of useful energy output to total energy input.

When using devices that are doing work it is important to know how much of the energy supplied is actually being converted into the required form for the job in hand, rather than wasted in the form of, for example, heat or sound.

The **efficiency** gives a quantitative measure of how much energy is changed into the correct form.

If we consider the energy output and input in one second we have, in effect, a power ratio:

$$\text{efficiency} = \frac{\text{useful energy output}}{\text{total energy input}}$$

or

$$\text{efficiency} = \frac{\text{useful power output}}{\text{total power input}}$$

To obtain the percentage efficiencies we simply multiply the fractional efficiency by 100.

For a large electric motor a typical efficiency would be 90%, whereas for a car engine it would be 30%. Any device that runs at a higher temperature than its surroundings will lose considerable heat energy to the surrounding medium and consequently have a lower efficiency.

Worked example

A hoist raises a mass of 480 kg through 9.0 m by means of an electric motor. The mass is raised at a constant speed of $1.2\,\mathrm{m\,s^{-1}}$.

(a) What is the gain in gravitational potential energy of the mass?

(b) How much mechanical power is the motor producing during the lift?

(c) If the motor is 90% efficient, how much electrical energy is used in the lift?

Answer

(a) $\Delta\mathrm{GPE} = 480 \times 9.81 \times 9.0 = 42\,379\,\mathrm{J}$

(b) $P = \text{rate at which energy is transferred} = \dfrac{\Delta\mathrm{GPE}}{t}$

$$\text{time} = \frac{\text{distance}}{\text{speed}} = \frac{9.0}{1.2} = 7.5\,\mathrm{s}$$

$$P = \frac{42\,379}{7.5} = 5651\,\mathrm{W}$$

(c) $\%\ \text{efficiency} = \dfrac{\text{useful power out}}{\text{total power in}} \times 100$

$$90 = \frac{5651}{\text{electric power supplied}} \times 100$$

$$P_{\text{electrical}} = \frac{5651}{90} \times 100 = 6278\,\mathrm{W}$$

P = rate at which energy is transferred

Therefore:

$$E_{\text{electrical}} = P_{\text{electrical}} \times \text{time} = 6278 \times 7.5 = 47\,088\,\mathrm{J}$$

(Note that 90% of the final answer equals the gain in gravitational potential energy.)

Examiner tip

We could have used $P = Fv$, knowing the force to be equal in magnitude to the weight:

$P = (480 \times 9.81) \times 1.2 = 5651\,\mathrm{W}$

Knowledge check 22

How much useful work is done by a horse in pulling a canal barge 120 m if the towrope makes an angle of 30° with the canal and the tension in the rope is 2200 N?

Knowledge check 23

With what speed will a can of beans hit the floor when it falls from a shelf 2.0 m above floor level?

Knowledge check 24

A lift and its occupants weigh 12 000 N. What is the minimum power output of the motor needed to raise the lift at a steady speed of $2.0\,\mathrm{m\,s^{-1}}$?

Knowledge check 25

A 5 W electric motor raises a weight of 15 N through a height of 2.0 m in 8.0 s. What is the efficiency of the motor?

Summary

- Work done is defined as the product of the force and the distance moved in the direction of the force.
- When work is done energy is transferred from one form into another.
- The change in the gravitational energy of a body is given by the equation: $\Delta\mathrm{GPE} = mg\Delta h$
- The change in the kinetic energy of a body is given by the equation: $\Delta\mathrm{KE} = \frac{1}{2}mv^2 - \frac{1}{2}mu^2$
- The principle of the conservation of energy states that energy cannot be created nor destroyed, but can be changed from one form into another.
- Power is defined as the rate at which work is done or energy is transferred. It is given by the equations:

  $P = \dfrac{\Delta E}{t}$

  $P = Fv$
- The efficiency of a machine is a measure of how much of the energy supplied is transferred into the desired form:

  $\text{efficiency} = \dfrac{\text{useful energy out}}{\text{total energy in}}$

  It can also be expressed as a power ratio.

Deformation of solids

Another consequence of the application of a force to a body is a change to its shape or dimensions — for example, it can be stretched, twisted or compressed. The change of shape is called deformation.

The specific nature of this deformation and the behaviour of materials under the action of forces provide vital information required by engineers in all applications, such as buildings, bridges and vehicle components.

Hooke's law

Hooke's law states that the extension of a body is directly proportional to the applied force up to a point known as the limit of proportionality.

By fixing a helical spring or a wire at one end and applying different stretching or tensile forces to the other, we can investigate the relationship between the size of the deforming force and the increase in length or extension it produces.

Robert Hooke summarised the results of such investigations. **Hooke's law** may be written in equation form:

$$F = kx$$

where:

- F is the deforming force
- x is the extension caused to the body by the force
- k is the force per unit extension or force constant

Up to the **elastic limit** the deformation will be temporary and when the deforming force is removed the body will return to its original state. The body is said to have behaved elastically. For some bodies the limit of proportionality will coincide with the elastic limit.

When the force is greater than that which caused the body to extend to the elastic limit the deformation will be permanent and the change is said to be **plastic**. It is usual with plastic deformation to see considerable extension, as the molecular structure breaks down, followed by fracture. This is best viewed in a graph of the experimental results obtained from stretching a wire (Figure 22).

Knowledge check 26

A suspended helical spring has an unstretched length of 20.0 cm. Its length is increased to 27.5 cm when a 6.0 N weight is hung from its lower end. What is the spring or force constant?

Knowledge check 27

What is the difference between elastic and plastic deformation?

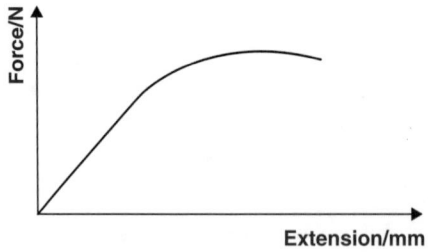

Figure 22

The slope of the straight line section in Figure 22, up to the limit of proportionality, will equal k, the force constant.

Worked example

A wire 2.40 m long is extended to a length of 2.47 m when a force of 67.0 N is applied.

(a) Assuming this force does not cause the wire to deform beyond the limit of proportionality, what is the force constant of the wire? State your unit.

(b) If a length of 2.50 m was needed, what tension would be required?

Answer

(a) extension, x = extended length – unstretched length

$x = 2.47 - 2.40 = 0.07\,\text{m}$

Hooke's law:

$F = kx$

Therefore:

$$k = \frac{F}{x} = \frac{67.0}{0.07} = 957\,\text{N m}^{-1}$$

(b) extension, x = extended length – unstretched length

$x = 2.50 - 2.40 = 0.10\,\text{m}$

Hooke's law:

$F = kx = 957 \times 0.10 = 95.7\,\text{N}$

Young modulus

If we were to change the thickness of the wire under investigation, the value of k, the force constant, would change. A thicker wire of the same material would need more force to cause the same extension.

Engineers need data for a material that are dimension free, so that they can directly compare materials to decide on their suitability for the task. The Young modulus is such a value — it gives a numerical value to the resistance of a material to deformation that is independent of its thickness or length.

The Young modulus is defined in terms of **stress** and **strain**.

Tensile stress will cause an increase in length and compressive stress leads to a decrease.

The unit of stress is newtons per metre squared, N m^{-2}, which is also the unit of pressure — called the pascal, P.

As strain is a **ratio** of two lengths it does not have a unit. The ratio of stress to strain is a constant for a material, provided the limit of proportionality has not been exceeded. This constant is known as the **Young modulus, *E***.

$$E = \frac{\text{stress}}{\text{strain}} = \frac{\sigma}{\varepsilon}$$

As there is no unit for strain, the unit of the Young modulus is the same as that for stress — the pascal. For example, the value for the Young modulus of copper is 1.1×10^{11} Pa.

When a force is applied to a material it will experience **stress**, σ, which is defined as force per unit area of cross-section:

$$\sigma = \frac{F}{A}$$

If stress is the cause then strain is the effect. **Strain**, ε, is defined as the ratio of the extension to the original length:

$$\text{strain, } \varepsilon = \frac{x}{L}$$

Knowledge check 28

Calculate the stress produced in a wire of diameter 0.20 mm when a tension of 40 N is applied.

Examiner tip

Note the parallel between E, the Young modulus for the material, and k, the force constant for the specific item of that material.

Examiner tip

- You should use a long length of wire — at least 2 m. To record the stretching of all the wire place a metre rule vertically as shown, with the zero of the ruler in line with the end of the unstretched, but straight, kink-free wire.
- Measure the original diameter of the wire in three different places with a micrometer screw gauge and use the average diameter in the equation $A = \frac{1}{4}\pi d^2$.
- Use a tape measure to find the original length from the clamped point to the end from which the masses are hung.

Experiment to determine the Young modulus

The Young modulus of a metal in the form of a thin wire can be measured, in a school laboratory, by carrying out the same experimental procedure that would be used to find the force constant for the wire — suspend known masses from the wire and measure the extension that results. The additional measurements required in order to find the Young modulus are the original length of wire being stretched and the diameter of the wire.

The typical arrangement used for thin copper is shown in Figure 23. The suspended masses provide a convenient source of known force. Each 100 g slotted mass provide approximately 1 N of tensile force.

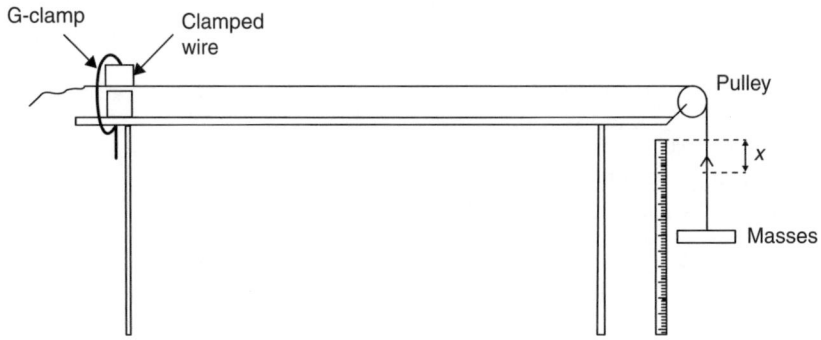

Figure 23

How to determine the Young modulus

Use the data to plot a graph of force, F, against extension, x.

From the linear section of the graph find the force constant, k, which equals the slope or gradient.

$$E = \frac{\text{stress}}{\text{strain}} = \frac{F/A}{x/L}$$

Rearranged:

$$E = \frac{F}{x} \times \frac{L}{A} = k\frac{L}{A}$$

Examiner tip

When calculating E in pascals, be sure to use SI numerical values for all quantities: force in N, x and L in m, and A in m^2.

Ultimate tensile stress

As the name suggests, this is the maximum stretching force the wire can support divided by the original cross-sectional area. It is not necessarily the same as the stress when a wire finally breaks, due to a phenomenon known as necking.

Summary

- Hooke's law states that the extension of a body is directly proportional to the applied force up to a point known as the limit of proportionality.
- The force constant, k, is the force per unit extension, up to the limit of proportionality.
- A material is said to be elastic if it regains its original dimensions when the deforming force is removed.
- A material is said to have undergone plastic deformation if it does not fully return to its original dimensions when the deforming force is removed.
- The Young modulus is constant for a material and is a measure of its resistance to deformation.
- Stress is defined as force per unit area.
- Strain is defined as extension per unit length.
- The Young modulus $= \frac{\text{stress}}{\text{strain}}$
- The unit of both stress and the Young modulus is the pascal, Pa.

Electricity

Electrical energy is classed as high-grade energy because it is easily transferred into other useful forms. It wasn't until the creation of the national grid in the 1930s that electricity became available to the majority of the population in England. Modern society is now totally reliant on electrical energy and therefore an understanding of the fundamental concepts is useful.

Current

For electrical energy to be moved from one place to another, where it can be converted to mostly heat and light, relies on electric current, which flows through conducting wires.

Electric current can be expressed as an equation:

$$I = \frac{\Delta Q}{\Delta t}$$

where I is the current and ΔQ is the amount of charge passing in time Δt.

The unit of current is the ampere, A.

The unit of charge is the coulomb, C.

Therefore:

1 ampere = 1 coulomb per second

In metal conductors the flowing charged particles are electrons, which are negatively charged.

The charge of a single electron, e, is 1.6×10^{-19} C.

$$I = \frac{\Delta Q}{\Delta t} = \frac{Ne}{\Delta t}$$

where N is the number of electrons. So when a current of 1 A is flowing in a wire, 1 C of charge is passing each point which means 6.25×10^{18} electrons pass each point in a second.

An **electric current** is defined as the rate at which charged particles pass a point in a circuit.

Examiner tip

Conventional current is denoted in circuit diagrams by an arrow in the direction from the positive terminal towards the negative terminal of the supply. Electrons move in the opposite direction.

Knowledge check 29

The current in a certain wire is 1.5 A. Calculate (a) the charge and (b) the number of electrons passing a point in the wire in 1 minute.

A **potential difference** is defined as the electrical energy converted per coulomb of charge passing between two points.

Potential difference

To make current flow between two points a **potential difference**, p.d., must exist between the two points. The defining equation is:

$$V = \frac{W}{Q}$$

where V = potential difference, W = work done or energy converted in the circuit and Q = charge in coulombs.

Potential difference is measured in volts and, as can be seen from the equation above:

1 volt = 1 joule per coulomb

Electrical power and energy

The defining equation for potential difference can be adapted to give a relationship that links electrical power, current and the potential difference.

Using the equations:

$$I = \frac{Q}{t} \text{ and } P = \frac{W}{t}$$

Rearranged:

$$Q = It \text{ and } W = Pt$$

Substituted:

$$V = \frac{W}{Q} = \frac{Pt}{It} = \frac{P}{I}$$

We get:

electrical power, $P = VI$

The unit for power, whether mechanical or electrical, is the watt, W.

Power, as stated, is the rate at which work is done or energy transferred:

$$P = \frac{W}{t}$$

Therefore the energy transferred can also be expressed by the equation:

energy transferred, $W = Pt = VIt$

The unit for all forms of energy is the joule, J.

Knowledge check 30

A 3 kW immersion heater is designed for use on the 240 V mains.

(a) What is the current taken from the mains when it is operating?

(b) What energy, in MJ, does it use in 15 minutes?

The **e.m.f.**, ε, of a source is the energy converted into electrical energy when unit charge passes through it.

Electromotive force

Batteries and generators supply energy to the charge as it passes through the source. The battery or generator is said to provide electromotive force, **e.m.f.**

The unit of e.m.f. is the volt. A car battery with an e.m.f. of 12 volts supplies 12 joules of energy per coulomb of charge passing through it.

Although e.m.f. and p.d. have the same unit, they are notably different. While the e.m.f. relates to a source and the energy it supplies, p.d. refers to electrical energy being converted in the circuit.

Examiner tip

The e.m.f. and the terminal p.d. will have the same numerical value when the current drawn from the cell is zero.

There is more about e.m.f. on pp. 50–52.

Electrical resistance

Charges experience opposition to their movement as they move through conductors. This is called **resistance**; it gives a numerical value to the opposition and is defined in terms of the potential difference needed and the current it produces.

Resistance may be measured between two points in a circuit but more commonly it is the resistance of a component in a circuit that is required.

$$\text{resistance, } R = \frac{V}{I}$$

The unit of resistance is the ohm, Ω, with 1 Ω being equivalent to 1 V A^{-1}.

A numerical value of the **resistance** is defined as the ratio of the potential difference between two points to the current passing through those points.

Knowledge check 31

The current flowing in a light bulb is 3.0 A when the p.d. is 12 V. What is the resistance of the bulb?

Summary

- Electric current is the rate of flow of charge:
  $$I = \frac{\Delta Q}{\Delta t}$$
- Potential difference measures the electrical energy transferred per coulomb of charge passing between two points:
  $$V = \frac{W}{Q}$$
- electrical power, $P = VI = I^2R = \frac{V^2}{R}$
- electrical energy transferred = VIt
- The e.m.f., ε, of a source is the energy converted into electrical energy when unit charge passes through it.
- $$\text{resistance, } R = \frac{\text{p.d. across a component}}{\text{current through it}} = \frac{V}{I}$$
- Resistance is dependent on the type and dimensions of the conductor.

Circuits

Conservation laws applied to circuits (Kirchhoff's laws)

Conservation of charge

Charge cannot physically leave a circuit, so the charge entering a junction, *per second*, is equal to the charge leaving the junction, *per second*. So the current into a junction is equal to the current out of the junction.

We can use the vector property of current here. Current in can be treated as positive and current out as negative, and so we can write:

$$I_1 + I_2 + I_3 = 0$$

See, for example, Figure 24.

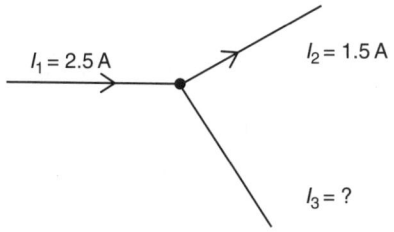

Figure 24

Knowledge check 32

For a wire that carries 0.4 A and has a resistance of 2 Ω, determine the total energy dissipated in 5 minutes.

Apply $I_1 + I_2 + I_3 = 0$:

$$2.5 + (-1.5) + I_3 = 0$$

Therefore $I_3 = -1.0\,\text{A}$, out of the junction.

Conservation of energy

Energy conservation will apply to electric circuits as it does everywhere else.

Charge will move from an energy source and follow a circuit loop, returning back to the energy source to be re-energised. As it passes round it will lose electrical energy to the components in the loop.

Imagine the journey of 1 coulomb of charge round a circuit loop. The total electrical energy converted at the components in the loop, which is equal to the sum of the p.d.s round the loop (as $V = \frac{W}{Q}$), is equal to the total energy supplied to it, which is equal to the e.m.f.:

energy in = energy delivered or transferred

energy in per coulomb = energy delivered or transferred per coulomb

$$\frac{E_{in}}{Q} = \frac{W}{Q}$$

e.m.f. = Σp.d.s

Series and parallel circuits

To show how components are electrically connected to the electrical energy source we draw circuit diagrams, using standard symbols. Components can either be connected in **series** or in **parallel**.

Examiner tip

Current arrows on wires point in the direction of 'conventional current', from the positive terminal towards the negative terminal. The direction of conventional current is opposite from the direction of electron movement.

In the **series** arrangement one component is followed by the next in a single path (Figure 25).

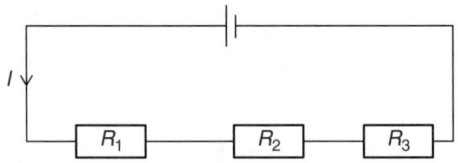

Figure 25

As a consequence the current is the same at all points in a series circuit and the potential differences across each component add up to the potential difference of the supply.

In a series circuit the total resistance is equal to the sum of the individual resistors:

$$R_{total} = R_1 + R_2 + R_3$$

In a **parallel** arrangement one component sits parallel to the next in a different loop; it is a multi-path circuit. The current will divide at a junction, but as no charge leaves the circuit the sum of the currents in will equal the sum of the currents out of a junction (Figure 26):

$$I = I_1 + I_2 + I_3$$

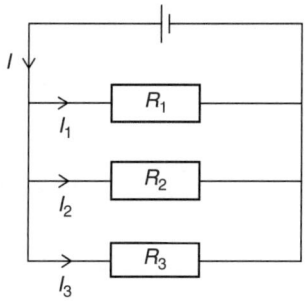

Figure 26

The potential difference across each branch of a parallel arrangement is equal.

For resistors connected in parallel, the total resistance is given by the equation:

$$\frac{1}{R_{total}} = \frac{1}{R_1} + \frac{1}{R_2} + \frac{1}{R_3}$$

Most working circuits will be a combination of series and parallel sections.

Worked example

In the circuit in Figure 27 B is a 13-volt battery of negligible internal resistance. The potential difference between X and Y is 5 V.

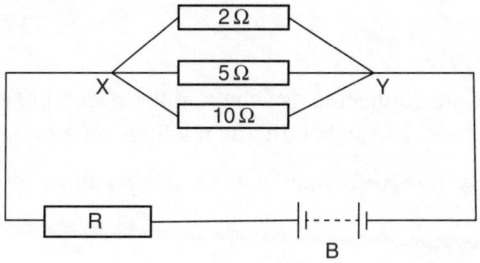

Figure 27

(a) Determine the effective resistance of the set of three parallel resistors between X and Y.

(b) What is the potential difference across the 2 Ω resistor?

(c) Determine the current flowing through the 10 Ω resistor.

(d) Determine the current flowing through the resistor R.

(e) Determine the magnitude of resistor R.

Examiner tip

Be sure to know the symbols for all components needed for the AS1 specification: battery, cell, variable power supply, switch, resistor, bulb, rheostat, thermistor, ammeter, voltmeter.

Knowledge check 33

What is the total resistance of three 12 Ω resistors placed in (a) series, (b) parallel?

Examiner tip
The effective resistance of resistors in parallel is always smaller than the smallest individual resistor.

Examiner tip
If two equal resistors are placed in parallel then the effective resistance of the combination is equal to half the value of the individual resistor. For example, for two 10 Ω resistors in parallel, the effective resistance is 5 Ω.

Answer

(a) $$\frac{1}{R_{\text{effective}}} = \frac{1}{R_1} + \frac{1}{R_2} + \frac{1}{R_3}$$

$$\frac{1}{R_{\text{effective}}} = \frac{1}{2} + \frac{1}{5} + \frac{1}{10} = \frac{5}{10} + \frac{2}{10} + \frac{1}{10} = \frac{8}{10}$$

Therefore:

$$R_{\text{effective}} = \frac{10}{8} = 1.25\,\Omega$$

(b) The p.d. across each branch in a parallel section is the same:

$$V_{XY} = 5\,\text{V} = V_{2\Omega}$$

(c) $$V_{XY} = 5\,\text{V} = V_{10\Omega}$$

$$R = \frac{V}{I} \text{ or } I = \frac{V}{R}$$

$$I_{10\Omega} = \frac{5}{10} = 0.5\,\text{A}$$

(d) Using the answers to (a) and (b) we determine the current entering XY.

$$I_{XY} = \frac{V_{XY}}{R_{\text{effective}}} = \frac{5}{1.25} = 4.0\,\text{A}$$

This is also the value of the current through resistor R.

(e) $$V_{\text{supply}} = V_R + V_{XY}$$

$$V_R = 13 - 5 = 8\,\text{V}$$

$$R = \frac{V_R}{I_R} = \frac{8}{4} = 2\,\Omega$$

Worked example

Two identical bulbs are connected as shown in the circuit in Figure 28.

(a) If ammeter A1 reads 0.6 A, what are the readings on A2 and A3?

(b) If bulb X is now removed, what will be the readings on each of the three ammeters?

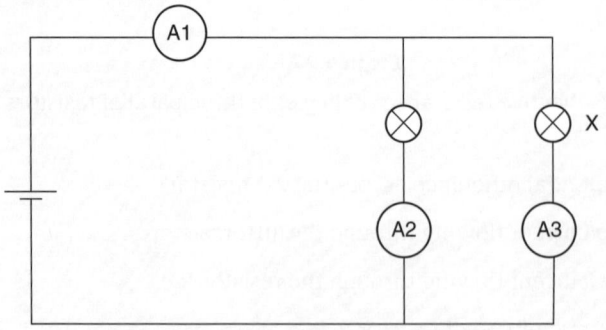

Figure 28

Answer

(a) Charge cannot be lost from a circuit; therefore there is conservation of charge and current at a junction:

$A1 = A2 + A3 = 0.6\,\text{A}$

As the bulbs have equal resistance the current through each parallel branch will be equal.

$A2 = A3 = 0.3\,\text{A}$

(b) With bulb X removed, equivalent to a break in the circuit, no current can take this path and $A3 = 0\,\text{A}$.
However the conditions in the other path have not changed from part (a); it is the same bulb with the same resistance and the p.d. across the bulb is unchanged. Consequently the ammeter $A2$ will read the same as before:

$A2 = 0.3\,\text{A}$.

Conservation of charge and thus current still applies.

$A1 = A2 + A3$

$A1 = 0.3 + 0.0 = 0.3\,\text{A}$

Examiner tip

This question is often answered incorrectly; it is worthwhile setting up the circuit to confirm the outcome. Common sense would also tell us that the current drawn in lighting one bulb should be half that for two.

Summary

- At a junction in a circuit the sum of the currents entering equals the sum of the currents leaving. This is a consequence of the law of the conservation of charge.
- $\Sigma I_{in} = \Sigma I_{out}$
- In a circuit loop the e.m.f. will equal the sum of the p.d.s.
- Conventional current direction is taken as from the positive terminal towards the negative terminal of the supply, although the electrons move in the opposite direction.
- The total resistance of any number of resistors in series is given by:
  $R_{total} = R_1 + R_2 + R_3$
- The total resistance of any number of resistors in parallel is given by:
  $\frac{1}{R_{total}} = \frac{1}{R_1} + \frac{1}{R_2} + \frac{1}{R_3}$

Current–voltage characteristics

A variety of components are used in electrical circuits, for example resistors, lamps, motors, thermistors, diodes and transistors. The behaviour or characteristics of these components is best displayed in a current–voltage graph.

The circuit in Figure 29 is used to test a component. By varying the supply p.d. a range of current and p.d. values can be recorded for the component using the ammeter and voltmeter.

Note that the supply can be reversed to get the negative p.d. values.

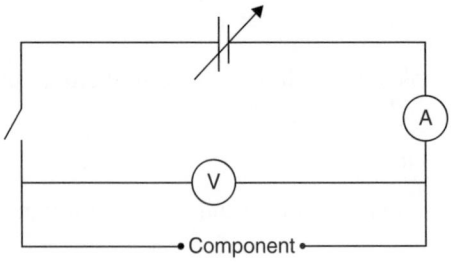

Figure 29

The results are displayed for each component in a current–voltage graph.

Figures 30–32 show the current–voltage characteristics for different components in a series of graphs.

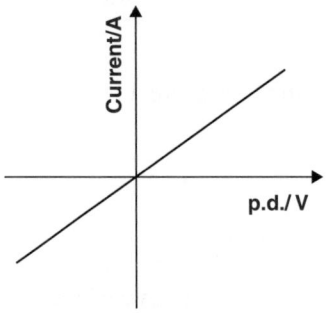

Figure 30 Metallic conductor at constant temperature

Figure 30 could be a wire resistor. The current and potential difference are directly proportional to each other when the current flows in either direction.

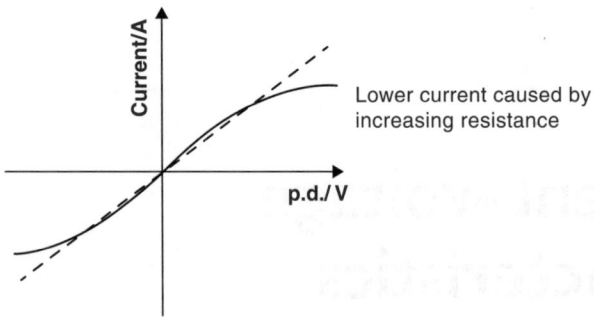

Figure 31 Metallic conductor *not* at a constant temperature.

Figure 31 could be the metal filament in a bulb. The current and potential difference are not proportional because the current heats the filament. An increase in the temperature of the wire increases the resistance of the filament.

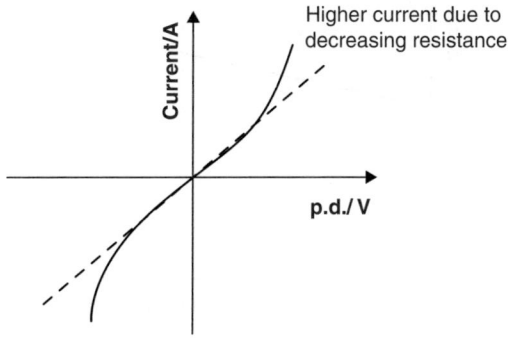

Figure 32 A negative temperature coefficient thermistor

The resistance of a thermistor changes rapidly with temperature. For a negative temperature coefficient thermistor semiconductor, the increase in temperature and with it thermal energy causes a significant increase in the number of conducting electrons and so an increased current value (Figure 32).

Ohm's law

The current–voltage characteristic shown in Figure 30 is referred to as ohmic behaviour and a conductor that displays this proportionality is an ohmic conductor Ohmic conductors are conductors that obey **Ohm's law**.

To identify a true non-ohmic conductor we need to see a current–voltage characteristic whose non-linear appearance is not a consequence of temperature change, for example a semiconductor diode.

Resistance and temperature

Electrical conductors have almost free, delocalised electrons, which flow under the influence of a potential difference. This is in contrast to insulators, where the electrons are tightly held to an individual atom and therefore there are no charge carriers and no current is possible.

A conductor also has vibrating positive ions. As the electrons flow they interact with these ions; it is this impedance to charge flow that constitutes resistance.

As the temperature of a metal increases, the positive ions vibrate with a greater amplitude. This results in greater opposition to electron flow and an increase in the resistance.

With a thermistor — a semiconductor — there is a second, more dominant effect. The increase in temperature also results in the release of more conduction electrons and therefore an overall decrease in the resistance.

This variation of resistance with temperature change can be confirmed experimentally and results displayed in graphical form (Figures 33 and 34).

Ohm's law states that the current through a metallic conductor is directly proportional to the potential difference across it, provided the temperature remains constant.

Examiner tip
Metals are ohmic conductors, yet the metal in a filament bulb — tungsten — does not display the proportionality. This is because the condition of having a constant temperature does not apply. The filament temperature will rise from room temperature to about 1500°C.

Examiner tip
Resistance should always be calculated from discrete values of *V* and *I*. It will only equal the inverse of the gradient of the *I*/*V* graph if this is a straight line.

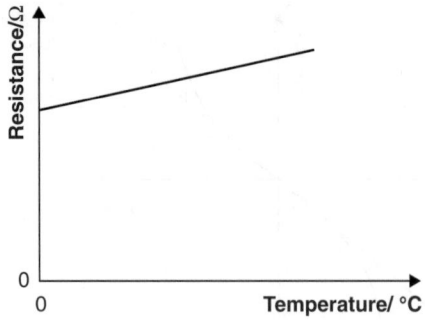

Figure 33 Resistance change of a conductor against temperature

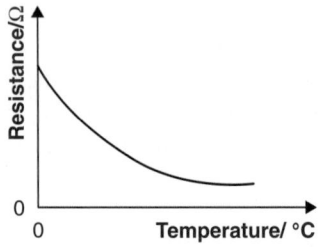

Figure 34 Resistance change of a semiconductor against temperature

Superconductivity

The resistance we have described in conductors due to the interaction between the electrons and the vibrating ions is seen to reduce as the temperature of the conductor becomes lower. If this temperature reduction is continued to temperatures near absolute zero then the vibration of the ions is almost zero and there should be little opposition to electron movement. (The cooling of the materials is achieved using liquid nitrogen or helium.)

Examiner tip
The critical temperature is also referred to as the transition temperature.

With further low temperature investigation it has been found that for several metals a zero resistance condition is achieved, not gradually but abruptly at a temperature now called the **critical temperature**, T_c. Materials that display this characteristic are called superconductors. Some ceramic materials have also been found to display the property of superconductivity — and at higher temperatures, for example at about 90 K.

Below the critical temperature the superconducting materials have no electrical resistance and so they can carry large electric currents without losing energy as heat (Figure 35).

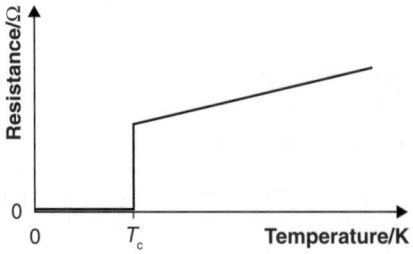

Figure 35 Resistance change of a superconductor against temperature

Superconductors that provide large currents with little energy loss are used in the powerful electromagnets required in MRI and accelerator technology.

Summary

- Ohm's law states that the current through a metallic conductor is directly proportional to the potential difference across it, provided the temperature remains constant.
- The resistance of a metallic conductor will increase as its temperature increases.
- The resistance of a negative temperature coefficient thermistor will decrease as the temperature of the thermistor increases. Thermistors are used as temperature sensors.
- A superconductor has zero resistance below a critical temperature.
- Superconductors are used to deliver the large currents needed to make high-power electromagnets.

Resistivity

The resistance of a wire at constant temperature depends on its dimensions, but also the material from which it is made. So there are three factors that determine the resistance of a material:

- length
- cross-sectional area
- type of material

Consider a wire:

- The longer the wire the further the electrons must go and the greater the number of interactions with positive ions, resulting in a resistance increase. Experiment confirms that the resistance of a wire is proportional to its length:

$$R \propto L$$

- The thicker the wire the greater the number of free electrons moving through the same distribution of positive ions, in effect a reduction in resistance. Experiment confirms the resistance of a wire is inversely proportional to its cross-sectional area:

$$R \propto \frac{1}{A}$$

- The degree of interaction will differ depending on the wire type. Copper is a good conductor while nichrome is a poor conductor.

Combining the first two experimentally confirmed relationships gives an overall equation for resistance, R:

$$R \propto \frac{L}{A} \text{ or } R = \text{constant} \times \frac{L}{A}$$

We have yet to include the third factor — a value to reflect wire type. This constant is a property of the material from which the wire is made, and is called its **resistivity**.

Resistivity is defined as being numerically equal to the resistance between opposite faces of a 1-metre cube of the material.

The resistivity, ρ, is a constant for a particular material, at constant temperature. It is a measure of the opposition to charge flow due to the nature or type of the material.

$$R = \text{constant} \times \frac{L}{A} = \frac{\rho L}{A}$$

Examiner tip

Resistivity is to resistance what the Young modulus is to the force constant. The resistance and the force constant will have different values for different wires of the same material, whereas the resistivity and Young modulus are constants for the material, no matter what its dimensions.

Knowledge check 34

Calculate the resistivity of the material used for a heating element of length 0.8 m with cross-sectional area 0.06 mm^2 and resistance 15 Ω.

Examiner tip

Stretch the wire first to remove any kinks, and then attach it along the length of a metre rule. Use a low voltage setting on the power pack to ensure that the current is low and the temperature variation minimal.

Examiner tip

Be careful with your units; it is best to work in metres throughout. When converting mm^2 to m^2, divide by 10^6.

Therefore:

$$\rho = \frac{RA}{L}$$

So the value of resistivity, ρ, is equal to R when $L = 1$ m and $A = 1$ m^2.

The units of resistivity, ρ, are Ω m.

The value of the resistivity of a material varies greatly, from an order of magnitude of 10^{-8} Ω m for a conductor to 10^{13} Ω m for an insulator.

Experimental determination of resistivity

Finding the resistivity of a wire involves the use of the two meter method — ammeter and voltmeter — to determine resistance, and measurement of the dimensions of the wire. In keeping with good experimental practice, a series of results should be obtained. For example, for a wire of fixed thickness, you might find the resistance of several lengths, from 30 cm to 100 cm.

The same circuit used for the current–voltage characteristic is suitable, with the wire placed in the component position (see Figure 29).

- Using a micrometer screw gauge, measure the diameter of the wire at three different positions and calculate the average diameter, $\langle d \rangle$. Use the equation:

  $$A = \frac{\pi \langle d \rangle^2}{4}$$

  to calculate cross-sectional area.
- Use:

  $$R = \frac{V}{I}$$

  to calculate the resistance of each length of wire used.
- Plot a graph of resistance, R, against the length, L.
- Find the gradient of the best-fit straight line for the points plotted.

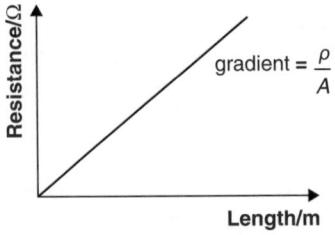

Figure 36

Using the equation:

$$R = \frac{\rho L}{A}$$

Rearranged:

$$\frac{\rho}{A} = \frac{R}{L} = \text{gradient}$$

Therefore:

$$\text{resistivity, } \rho = \text{gradient} \times A$$

Worked example

The heater for the rear window of a car consists of two strips of good electrical conductor joined by six thin strips of resistive metal (Figure 37). Each thin strip is 0.80 m long, with cross-sectional area 0.12 mm^2 and is made from material of resistivity $4.5 \times 10^{-6}\,\Omega\,\text{m}$ at room temperature.

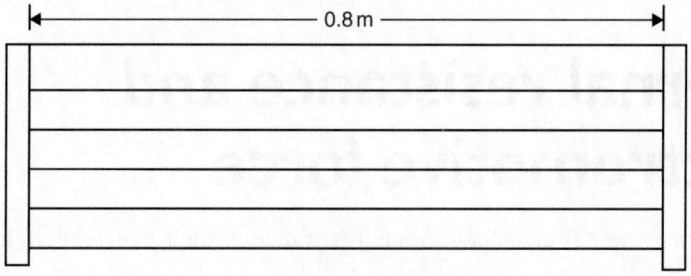

Figure 37

(a) Calculate the resistance of one of the strips at room temperature.

(b) Calculate the resistance of the complete heater at room temperature.

(c) Taking the terminal p.d. of the car battery to be 12 V, what power is delivered by the battery to the heater?

(d) Why are the strips connected as shown?

Answer

(a) $R = \dfrac{\rho L}{A}$

$$R = \frac{(4.5 \times 10^{-6}) \times 0.8}{0.12 \times 10^{-6}}$$

$$R = 30\,\Omega$$

(b) Six *equal* resistors in parallel:

$$\frac{1}{R_{\text{total}}} = \frac{1}{R_1} + \frac{1}{R_2} + \frac{1}{R_3} + \frac{1}{R_4} + \frac{1}{R_5} + \frac{1}{R_6}$$

$$\frac{1}{R_{\text{total}}} = \frac{6}{R}$$

$$\frac{1}{R_{\text{total}}} = \frac{6}{30} \text{ or } R_{\text{total}} = 5\,\Omega$$

(c) $P = VI$ or I^2R or $\dfrac{V^2}{R}$

$$P = \frac{12^2}{5} = 28.8\,\text{W}$$

(d) Connected in parallel so that if one strip breaks, the others continue to operate.

Summary

- Resistivity is a measure of a material's opposition to current and, unlike resistance, is independent of the dimensions of the sample.
- Resistivity, $\rho = \frac{RA}{L}$
- The unit of resistivity is the Ωm.

Internal resistance and electromotive force

All electrical power sources have some resistance of their own, called the internal resistance.

Each charge gains energy as it moves through the source, but some of the energy is used up in pushing the charge through the source. This energy is dissipated as heat, which is why a battery can become warm when supplying current.

We can treat the internal resistance, r, as a small resistor in series with the supply (Figure 38).

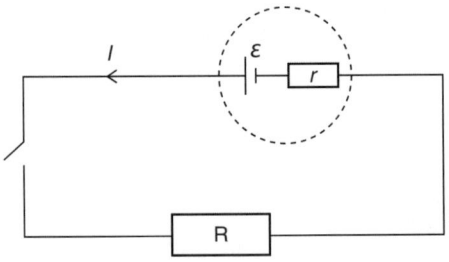

Figure 38

$V_{terminal}$ is the terminal potential difference, the p.d. between the terminals of a battery or cell when current is being delivered.

For a supply with e.m.f. ε and internal resistance r, delivering current I to a circuit of resistance R:

$$\varepsilon = IR + Ir$$

$IR = V_{terminal}$, the potential difference available to the circuit.

$Ir = V_{lost}$, the potential difference used to push the charge through the cell.

The equation can be written in different forms:

$$\varepsilon = V_{terminal} + V_{lost}$$

$$\varepsilon = V_{terminal} + Ir$$

$$\varepsilon = IR + V_{lost}$$

The effects of the internal resistance:

- When $I = 0$ A, apply $\varepsilon = V_{terminal} + Ir$; $\varepsilon = V_{terminal}$
- When I is high, apply $\varepsilon = V_{terminal} + Ir$; $V_{terminal}$ will be low.

Worked example

What is the terminal p.d. of a car battery of e.m.f. 12 V and r 0.04 Ω, when it is delivering 100 A to the starter motor?

Answer

$$\varepsilon = V_{\text{terminal}} + Ir$$

$$V_{\text{terminal}} = \varepsilon - Ir = 12 - (100 \times 0.04) = 8\,\text{V}$$

(It is this drop in terminal p.d. that is responsible for the lights of a car dimming as the engine is turned on.)

Examiner tip

When recharging a battery the potential difference applied across its terminals must be greater than the e.m.f.

Experimental determination of the internal resistance

The circuit in Figure 39 can be used to measure the e.m.f. and the internal resistance of a cell.

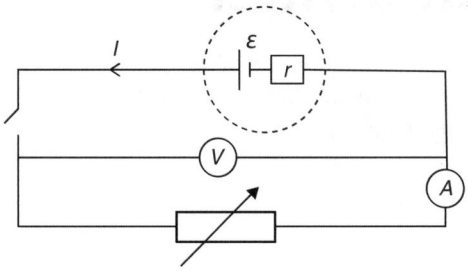

Figure 39

- Adjust the variable resistor to its maximum resistance value.
- Then close the switch and immediately record the values of the terminal p.d. and current from the voltmeter and ammeter respectively.
- Reopen the switch and decrease the resistance of the variable resistor.
- Close the switch again and record the new values on the voltmeter and ammeter.
- Repeat this process several times and then plot the terminal p.d. on the y-axis against current on the x-axis (Figure 40).

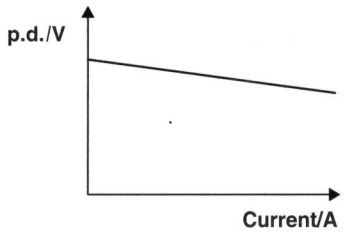

Figure 40

Knowledge check 35

Under what conditions is the e.m.f. of a supply:

(a) greater than the terminal p.d.

(b) equal to the terminal p.d.

(c) less than the terminal p.d.

Compare the equation for the e.m.f. and the general linear graph equation:

$V_{terminal} = -Ir + \varepsilon$ and $y = mx + c$

y-axis intercept, $c = \varepsilon$, e.m.f.

gradient, $m = -r$, internal resistance

Summary

- For a supply with e.m.f. ε and internal resistance r, $V = \varepsilon - Ir$, where V is the terminal potential difference available to the circuit.
- The terminal p.d. will always be less than the e.m.f. when the supply is delivering current (due to internal resistance).
- Dry cells have a significant internal resistance, whereas a car battery must have a low internal resistance as it needs to produce a large current for the starter motor.

Potential divider

There are occasions when only a fraction of the p.d. provided by a source is required, so we need to tap part of the p.d. available. This can be achieved by using two resistors in series with the supply (Figure 41).

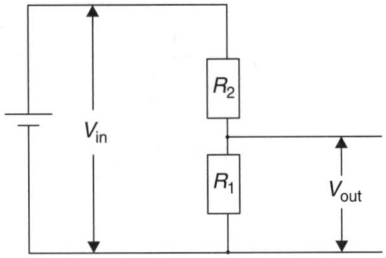

Figure 41

$$V_{in} = I(R_1 + R_2)$$

$$V_{out} = IR_1$$

where I is the current through the two resistors in series with the supply.

Combining the two equations:

$$\frac{V_{in}}{(R_1 + R_2)} = I = \frac{V_{out}}{R_1}$$

Rearranged:

$$\frac{V_{out}}{V_{in}} = \frac{R_1}{(R_1 + R_2)}$$

For a two-resistor potential divider:

$$V_{out} = \frac{R_1 V_{in}}{(R_1 + R_2)}$$

Examiner tip

When using the equation in this form, note the relative positions of R_1 and R_2. It is best to be able to generate the equation from first principles.

Variable potential

On occasions we may want a continuous range of p.d.s, for example in a school variable power supply or in a lighting dimmer circuit. This is easily obtained by replacing the two discrete resistors R_1 and R_2 by one slide variable resistor. Examples are illustrated in Figure 42.

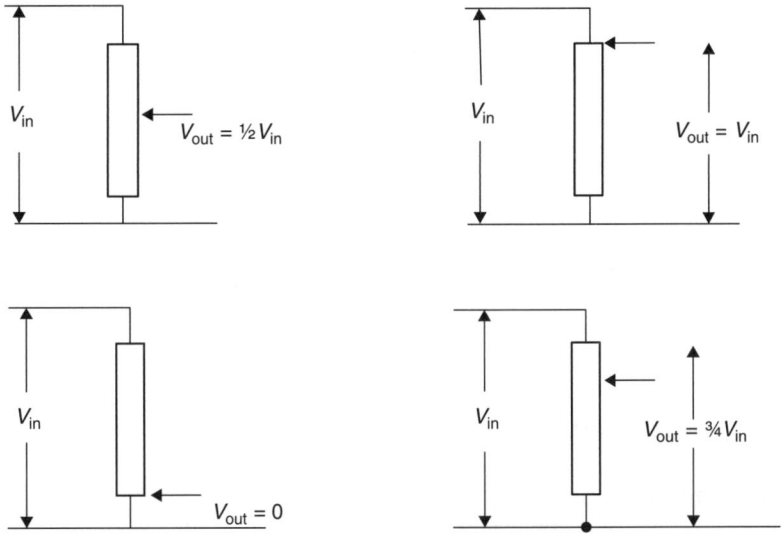

Figure 42

Examiner tip

This circuit has the advantage over simply placing a light in series with the variable resistor in that zero p.d. can be achieved with the potential divider but not with the simpler circuit.

Examiner tip

When a variable resistor is used to control current in a circuit, we refer to it as a rheostat. When a variable resistor is used to control voltage in a circuit, we refer to it as a potential divider.

Loading the potential divider

The component or load is placed across the V_{out} terminals, for example a bulb (Figure 43).

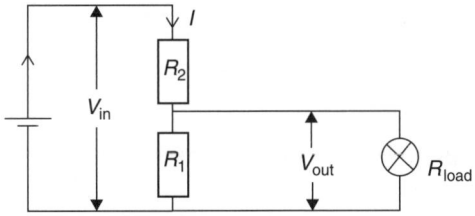

Figure 43

The V_{out} that will be predicted by use of the equation is theoretical.

Placing a load in parallel with R_1 will reduce the effective resistance of this part of the circuit and consequently the p.d.

Before it is applied, the effective resistance of R_1 and R_{load} should be calculated and used as a replacement for the R_1 value in the equation:

$$\frac{1}{R_{eff}} = \frac{1}{R_1} + \frac{1}{R_{load}}$$

$$V_{out} = \frac{R_{eff}V_{in}}{(R_{eff} + R_2)}$$

Worked example

(a) Calculate the potential difference between X and Y in Figure 44.

(b) Calculate the potential difference between X and Y when *another* 4 Ω resistor is connected between these points.

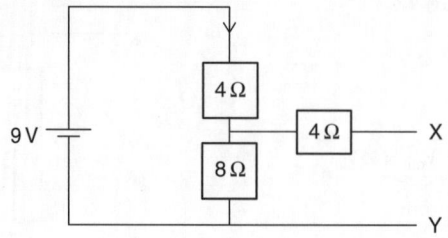

Figure 44

Answer

(a) $V_{out} = \dfrac{R_1 V_{in}}{(R_1 + R_2)}$

$V_{XY} = \dfrac{8 \times 9}{(4 + 8)} = 6\,\text{V}$

(b) See Figure 45.

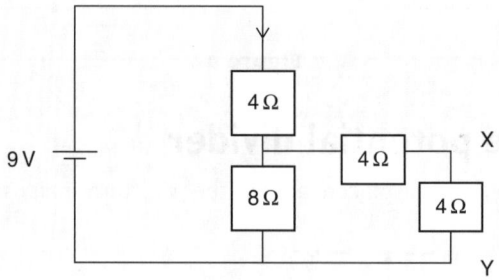

Figure 45

Two 4 Ω resistors in series, $4 + 4 = 8\,\Omega = R_{load}$.

8 Ω in parallel with R_1, of value 8 Ω.

$$\frac{1}{R_{eff}} = \frac{1}{R_1} + \frac{1}{R_{load}}$$

$$\frac{1}{R_{eff}} = \frac{1}{8} + \frac{1}{8} = \frac{2}{8} = \frac{1}{4}$$

$$R_{eff} = 4\,\Omega$$

$$V_{out} = \frac{R_{eff} V_{in}}{(R_{eff} + R_2)}$$

$$V_{XY} = \frac{4 \times 9}{(4 + 4)} = 4.5\,\text{V}$$

Sensors as part of the potential divider circuit

Thermistors and light-dependent resistors, LDRs, can be incorporated into potential dividers to cause a variation in the output voltage, which in turn can be used to trigger the operation of lighting or heating circuits. (The second resistor can be a variable resistor, so acting as a sensitivity control.)

Examiner tip

The LDR has a low resistance in bright light and a high resistance in darkness.

Heater circuit

One of the two resistors in the potential divider is replaced by a thermistor. As the temperature surrounding the thermistor falls, its resistance increases and the V_{out} increases to a value needed to switch on the heater (Figure 46).

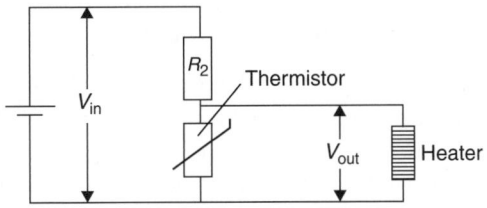

Figure 46

Lighting circuit

One of the two resistors in the potential divider is replaced by an LDR. As the light intensity surrounding the LDR falls its resistance increases and the V_{out} increases to a value needed to switch on a light (Figure 47).

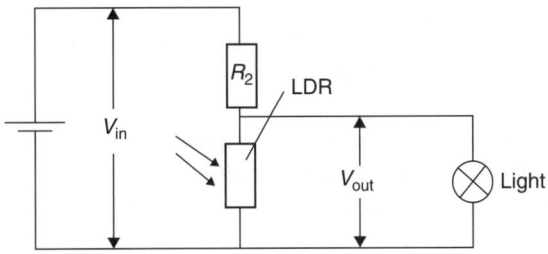

Figure 47

Summary

- Two or more resistors connected in series to a supply act as a potential divider.
- For a two-resistor potential divider:

$$V_{out} = \frac{R_1 V_{in}}{(R_1 + R_2)}$$

- The potential difference output of a potential divider circuit will drop once it is connected to a load.
- Variable potential dividers can be used in volume control and dimmer circuits, to allow variation from zero to a maximum.
- Variable potential dividers can be used in sensor circuits to activate lighting and heating.

Questions & Answers

The unit assessment

Unit AS1 is a written examination consisting of a number of compulsory short-answer questions. Some of the questions may require extended responses and most will contain a mathematical calculation component.

The maximum mark for the paper is 75 and the duration of the examination is 1 hour 30 minutes. AS1 contributes 37% to the overall AS award.

Assessment in AS qualifications requires students to demonstrate their quality of written communication. In particular you must ensure that text is legible and that spelling, punctuation and grammar are accurate, while the meaning is clear and relevant. Select a form of writing appropriate to the purpose, in a logical sequence, and include scientific vocabulary where appropriate.

Command terms

Examiners use certain words to indicate the type of response expected. (The depth of response required is usually indicated by the number of marks allocated.) It is helpful to be familiar with these terms.

- **Define** — a formal statement. If an equation is given, each term used must be identified.
- **State** — a concise answer with little or no supporting argument.
- **Explain** — usually a definition together with some relevant comment on the significance of the term.
- **List** — a number of points or terms with no explanation.
- **Describe** — state the key points relating to a particular concept or experiment.
- **Calculate** — used when a numerical answer is required. Working out should be shown.
- **Measure** — indicates that a quantity can be obtained directly by use of a suitable piece of apparatus.
- **Determine** — indicates that a quantity cannot be obtained directly but rather by calculation, with given or measured values substituted in a suitable form of a known equation.
- **Show** — used when a calculation has to be performed to obtain an already given value. All work must be shown clearly and logically, with the final value shown to a greater significance than required to confirm that it has not been a fudge.
- **Sketch** — when related to graphs, you are expected to show the general shape or trend of the graph, including intercepts if appropriate. The detail on the axes — quantities, units, origin — is normally expected. Additional values will be specifically asked for if required.
- **Draw** — a carefully and fully labelled diagram, with apparatus shown, using standard symbols, to enable all necessary measurements to be taken. Use a ruler where appropriate and do not make it small. Note the difference between a circuit diagram and a drawing of the apparatus required.

Symbols, signs and abbreviations

Questions will be set using SI units. Students must be familiar with the symbols for quantities relevant to the specification and with the units for these quantities. Questions will assume knowledge of the symbols for the decimal multiples and sub-multiples shown in the table below.

Multiplying factor	Prefix	Symbol	Multiplying factor	Prefix	Symbol
10^{-2}	centi	c	10^{3}	kilo	k
10^{-3}	milli	m	10^{6}	mega	M
10^{-6}	micro	μ	10^{9}	giga	G
10^{-9}	nano	n	10^{12}	tera	T
10^{-12}	pico	p			
10^{-15}	femto	f			

Data and Formulae Sheet for AS1

There will be a data and formulae sheet provided inside the AS1 examination paper. The constants given may be required to allow you to complete some of the calculations. Note that all the values are given to 3 significant figures.

The following values are provided:

- speed of light in a vacuum $c = 3.00 \times 10^{8}\,\mathrm{m\,s^{-1}}$
- elementary charge $e = 1.60 \times 10^{-19}\,\mathrm{C}$
- the Planck constant $h = 6.63 \times 10^{-34}\,\mathrm{J\,s}$
- mass of electron $m_e = 9.11 \times 10^{-31}\,\mathrm{kg}$
- mass of proton $m_p = 1.67 \times 10^{-27}\,\mathrm{kg}$
- acceleration of free fall on the Earth's surface $g = 9.81\,\mathrm{m\,s^{-2}}$
- electron volt $1\,\mathrm{eV} = 1.60 \times 10^{-19}\,\mathrm{J}$

The following relevant formulae are provided:

- conservation of energy $\frac{1}{2}mv^2 - \frac{1}{2}mu^2 = Fs$ for a constant force
- Hooke's law $F = kx$ (spring constant, k)
- terminal potential difference $V = \varepsilon - Ir$ (e.m.f., ε; internal resistance, r)
- potential divider $V_{out} = \dfrac{R_1 V_{in}}{R_1 + R_2}$

Revision tips

- The examination is designed to test all the content of the unit, and as all the questions are compulsory, your revision must address every element.
- Break down and learn in detail the content stated in the specification:
  - experiment descriptions (six named in AS1)
  - full definitions
  - equations, in all forms
  - statements of laws and principles
- Sequence your notes and corrected example questions in the order of the specification.
- Use past paper questions and those provided in this guide to ensure that you are familiar with typical questions associated with each topic.

About this section

This section consists of two self-assessment tests, each of 75 marks in total, as in the real exam. Allow 1½ hours if completing one of these tests in one sitting.

Try the questions without reference to the answers, then compare your responses with the given solutions and read the examiner's comments to give you an added insight to each problem and the probable pitfalls.

For question parts worth multiple marks, ticks (✓) are included in the answers to indicate where the examiner has awarded marks.

Examiner comments

Examiner comments on some questions are preceded by the icon e. They offer tips on what you need to do to gain full marks. Some answers are followed by examiner comments, indicated by the icon e, which highlight where credit is due or could be missed.

Self-assessment test 1

Question 1

(a) Show that the base units of potential energy and kinetic energy are the same. (4 marks)

Be clear of the difference between the SI unit for the quantity and the base units. The SI unit for any form of energy is the joule. Note that the SI unit of mass is the kilogram and not the gram.

(b) A mass *m* is attached to one end of a suspended spiral spring. This produces a force *F* in the spring. The mass is then pulled down and released, causing it to oscillate. Equation 1 represents the relationship for the periodic time *T* of mass–spring oscillation.

$$T = 2\pi\sqrt{\frac{my}{F}}$$ **Equation 1**

Determine the base units of the term *y*. (2 marks)

Answer to Question 1

(a) potential energy = mgh ✔
where g is the acceleration due to gravity.
units of mass = kg
units of acceleration = $m\,s^{-2}$
units of height = m
base units of potential energy = (kg) × ($m\,s^{-2}$) × (m) = $kg\,m^2\,s^{-2}$ ✔
kinetic energy = $\frac{1}{2}mv^2$ ✔
units of mass = kg
units of velocity = $m\,s^{-1}$
base units of kinetic energy = (kg) × $(m\,s^{-1})^2$ = (kg) × ($m\,s^{-1}$) × ($m\,s^{-1}$) = $kg\,m^2\,s^{-2}$ ✔
Both forms of energy have the base units $kg\,m^2\,s^{-2}$.

Make sure you show all stages of the solution to convince the examiner that your answer is not a bluff.

(b) base units of periodic time T = s
base units of mass m = kg
base units of force F = $kg\,m\,s^{-2}$ ✔

It is often simpler to remove a square root by squaring both sides of the equation.

$$T^2 = 4\pi^2 \frac{my}{F}$$

Rearrange the equation with y as the subject:

$$y = \frac{T^2F}{4\pi^2 m}$$

Enter the base units of the quantities:

$$\text{base units of } y = \frac{(\text{s}^2) \times (\text{kg m s}^{-2})}{(\text{kg})} = \text{m} \checkmark$$

In general when a derived unit is involved in a question — for example power, W — be sure to be working in the SI base units for mass, length, time and temperature.

Question 2

(a) State the difference between a scalar and a vector quantity. (1 mark)

(b) There are four physical quantities listed below. Indicate which of the quantities are scalars by underlining them. (2 marks)

velocity frequency kinetic energy density

(c) A ball is kicked towards a wall with a velocity of 3.5 m s^{-1} and rebounds directly back at the same speed. What is the change in the velocity of the ball? (2 marks)

When asked about a difference, always make reference to both quantities.

Answer to Question 2

(a) Vector quantities have a directional aspect; scalars do not. ✔

(b) <u>frequency</u> <u>kinetic energy</u> <u>density</u> ✔✔ (½ each, round up; penalty of 1 mark if velocity given)

(c) Whilst speed has not changed, the velocity after hitting the wall is –3.5 m s^{-1}. ✔
change in velocity = velocity after rebound – velocity before
change in velocity = (–3.5) – (3.5) = –7.0 m s^{-1} ✔

Note that the negative in the answer would not be a marking point on this occasion as only the *change* was asked for. The answer should be quoted to the same number of significant figures as the data given.

Question 3

The diagram below shows a force of 11 N acting on a block resting on a horizontal surface.

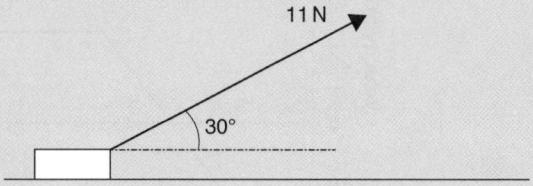

(a) Find the horizontal and vertical components of the force. (2 marks)

(b) What is the resultant vertical force acting on the horizontal surface if the block has a mass of 4.0 kg? (2 marks)

Answer to Question 3

(a) horizontal component of the force = $11 \cos 30° = 9.5\,\text{N}$ ✔

vertical component of the force = $11 \sin 30° = 5.5\,\text{N}$ ✔

(b) weight = $mg = 4.0 \times 9.81 = 39.2\,\text{N}$ ✔

resultant force = weight – vertical component of the force

resultant force = 39.2 – 5.5 = 33.7 N or 34 N ✔

ⓔ Use the value shown on the calculator from earlier question parts and not rounded values when substituting into a later equation. Then carry out one final rounding to the appropriate significance at the end (in this case two).

Question 4

A car is stopped at traffic lights. The lights change to green and just as the car starts to move, a bus, in the bus lane, passes it moving at a steady velocity of 15.0 m s^{-1}. The car accelerates at a steady rate, reaching a velocity of 22.5 m s^{-1} in 30 s and then maintains this velocity.

(a) Taking it that the lights become green at time = 0, draw the velocity–time graphs for the car and bus up to t = 60 s. Use the same axes for both. (3 marks)

(b) How long does it take the car to reach the same velocity as the bus? (1 mark)

(c) At that time, what is the distance between the car and the bus? (2 marks)

(d) What is the acceleration of the car in the time interval 0–30 s? (2 marks)

(e) How long after the lights change does the car pass the bus? (3 marks)

Answer to Question 4

(a)

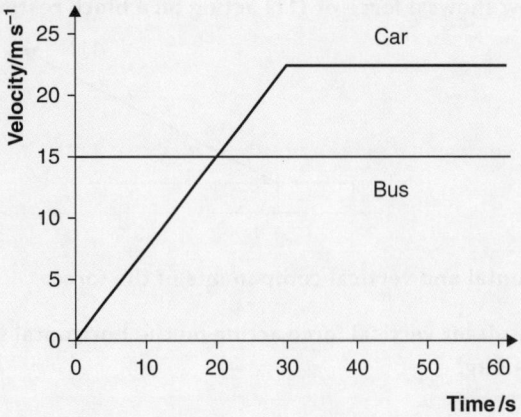

graph/scales ✔
bus — horizontal line ✔
car — diagonal and horizontal ✔

The only odd number that is acceptable to use in scaling is 5. Be sure to give the units and label in the correct format, quantity/unit, to give a pure number. Graphs only deal with pure numbers.

(b) intersection of the two plots at 20 s ✔

(c) area between the graphs and the time axis up to 20 s gives the distance travelled ✔
for the bus: $20 \times 15 = 300\,\text{m}$
for the car: $\frac{1}{2}(20 \times 15) = 150\,\text{m}$
150 m apart ✔

(d) Acceleration is the slope of the velocity–time plot. ✔

$$a = \frac{\Delta v}{\Delta t} = \frac{22.5}{30} = 0.75\,\text{m s}^{-2}$$ ✔

(e) Find the distance travelled by both after 30 s.
bus: distance = $30 \times 15 = 450\,\text{m}$
car: distance = $\frac{1}{2}(30 \times 22.5) = 337.5\,\text{m}$
112.5 m apart ✔
From now on the car catches up on the bus by $(22.5 - 15) = 7.5\,\text{m}$ each second. ✔
Therefore it will take a further $(112.5/7.5) = 15\,\text{s}$ to pass the bus.
This is 45.0 s after the lights changed. ✔

(e) is an unstructured, multi-step question. It is important to show as much of each step as you can, so if you know something about the latter stage only, show it. Error carried forward can be applied within a multi-step answer.

Question 5

An aircraft at a height of 1.2×10^4 m is travelling horizontally at 360 km h^{-1}. When the aircraft is vertically above a target, a bomb is released.

(a) How long before the bomb hits the ground? (2 marks)

(b) How far from the target does the bomb hit the ground? (4 marks)

Answer to Question 5

(a) The bomb has no initial vertical component of velocity.

Use $s = ut + \frac{1}{2}at^2$: ✔

$$t^2 = \frac{2s}{a} = \frac{2 \times 1.2 \times 10^4}{9.81} = 2446$$

$t = 49.5$ s ✔

(b) Convert km h^{-1} to m s^{-1}:

$$\frac{360 \times 1000}{60 \times 60} = 100 \text{ m s}^{-1}$$ ✔

The bomb has a horizontal velocity equal to that of the aircraft. ✔

There is no horizontal acceleration. ✔

Use $s = ut + \frac{1}{2}at^2$; reduces to $s = ut$:

$s = (100 \times 49.5) = 4950$ m ✔

Note that 50% of the marks in part (b) are for comments on the principles, without calculation. This again highlights that it is worth recording all you know pertaining to the question.

Question 6

(a) State Newton's second law of motion. (2 marks)

(b) When a car engine is providing a driving force of 700 N it moves along a level road with an acceleration of 0.4 m s^{-2}.
The driving force is then increased to 1200 N and the new acceleration is 0.8 m s^{-2}.
Assuming the frictional forces on the car remain constant:

(i) Calculate the value of the frictional force. (4 marks)

(ii) Determine the mass of the car. (1 mark)

Always read the full question. The overall context and specific clues to the relevant equations will appear in the wording, in this case the forces involved — driving and friction.

Answer to Question 6

(a) For a body of constant mass, its acceleration is directly proportional to the resultant force applied to it ✔ and in the direction of the resultant force ✔.

Newton's second law can be expressed in terms of 'rate of change of momentum', but the version given here is more in line with the specification. Minor word variations are acceptable, for example net force rather than resultant force.

(b) (i) $F_{resultant} = ma$ and $F_{resultant} = F_{driving} - F_{friction}$

Therefore:

$ma = F_{resultant} = F_{driving} - F_{friction}$ ✔

Apply to the two situations:

$m \times 0.4 = (700 - F_{friction})$ (Equation A) ✔

$m \times 0.8 = (1200 - F_{friction})$ (Equation B) ✔

If we multiply Equation A by 2 we can substitute this into Equation B:

$2(m \times 0.4) = 0.8m = 2(700 - F_{friction})$

$0.8m = (1200 - F_{friction}) = 2(700 - F_{friction})$

Therefore:

$1200 - F_{friction} = 1400 - 2F_{friction}$

$F_{friction} = 200\,\text{N}$ ✔

(ii) Substitute the value for friction into Equation A to determine the mass:

$m \times 0.4 = (700 - F_{friction}) = 700 - 200 = 500$

$m = 1250\,\text{kg}$ ✔

You can quickly check your answers by confirming that the substitution into Equation B gives the same answer.

Question 7

A 2 kW kettle is half filled with water and plugged into a 230 V mains supply. After 3 minutes the water is boiling and the kettle automatically switches off.

(a) (i) Calculate the heat energy dissipated by the element in the 3 minutes. (1 mark)

(ii) Calculate the amount of charge that passes through the element of the kettle in this time. (2 marks)

(iii) Hence calculate the number of electrons that pass through the element in this time. (2 marks)

(b) Calculate the resistance of the element. (2 marks)

Answer to Question 7

(a) (i) $W = P \times t = (2000 \times 180) = 360\,000\,\text{J}$ ✔

(ii) $W = QV$ ✔

and

$$Q = \frac{W}{V} = \frac{360\,000}{230} = 1565\,\text{C}$$ ✔

Alternatively:

$$P = VI \text{ or } I = \frac{P}{V} = \frac{2000}{230} = 8.7\,\text{A} \checkmark$$

$$Q = It = 8.7 \times 180 = 1565\,\text{C} \checkmark$$

Always remember to convert time into the base unit, seconds.

(iii) $Q = N \times e$ ✔

$$N = \frac{Q}{e}$$

N = number of electrons

e = charge of one electron = 1.6×10^{-19} C

$$N = \frac{1565}{(1.6 \times 10^{-19})} = 9.78 \times 10^{21} \checkmark$$

(b) $$R = \frac{V}{I} = \frac{230}{8.7} = 26.4\,\Omega \checkmark\checkmark$$

Alternatively:

$$P = \frac{V^2}{R}, \text{ so } R = \frac{V^2}{P} \checkmark$$

$$R = 230^2/2000 = 26.5\,\Omega \checkmark$$

The value of R obtained is slightly different owing to using the rounded value of I. Either answer is acceptable.

Question 8

(a) A piece of wire is stretched so that its length is increased by 1%, but its volume remains unchanged.

(i) How will the resistivity of the wire be affected, if at all? (1 mark)

(ii) How will the resistance of the wire be affected, if at all? (2 marks)

(b) The circuit below shows two different 6.0 V bulbs powered by the same battery. Bulb A is rated as 2.4 W and bulb B as 4.8 W.

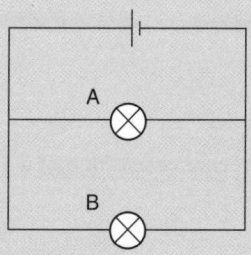

(i) Calculate the current delivered by the battery. (2 marks)

(ii) Calculate the resistance of the filament of bulb A. (2 marks)

The filament of bulb A is constructed from wire of cross-sectional area $2.0 \times 10^{-8}\,\text{m}^2$ and is made of material of resistivity $7.5 \times 10^{-6}\,\Omega\text{m}$.

(iii) Use your answer in (ii) to calculate the length of the filament of bulb A. (2 marks)

(iv) Suggest how this length of wire may be fitted into a small light bulb. (1 mark)

Answer to Question 8

(a) **(i)** Resistivity is a property of the material, which has not changed, so the resistivity is unchanged. ✔

(ii) Resistance is dependent on the dimensions of the wire, which have changed, so the resistance will change. ✔

$$R = \frac{\rho L}{A}$$

L increased by 1%. A will decrease by 1% as volume is constant and $V = L \times A$. Consequently R will increase by 2%; thinner and longer. ✔

(b) **(i)** $P_{total} = P_A + P_B = 2.4 + 4.8 = 7.2\,\text{W}$ ✔

$$P = VI \text{ or } I = \frac{P}{V} = \frac{7.2}{6} = 1.2\,\text{A} \checkmark$$

Alternatively you can work out the current for each bulb and add them:

$$I_A = \frac{P}{V} = \frac{2.4}{6.0} = 0.4\,\text{A} \checkmark$$

$$I_B = \frac{P}{V} = \frac{4.8}{6.0} = 0.8\,\text{A}$$

$$I_{total} = I_A + I_B = 0.4 + 0.8 = 1.2\,\text{A} \checkmark$$

(ii) $R = \frac{V}{I}$ ✔

$$R_A = \frac{6.0}{0.4} = 15\,\Omega \checkmark$$

(iii) $R = \frac{\rho L}{A}$ ✔

Rearranged:

$$L = \frac{RA}{\rho} = \frac{(15 \times 2.0 \times 10^{-8})}{7.5 \times 10^{-6}} = 4 \times 10^{-2}\,\text{m or } 4\,\text{cm} \checkmark$$

(iv) Coiled, but not touching. ✔

Check your specification; if it refers to 'recall and use', a mark may be awarded for simply recording the equation, as it will not appear on the *Data and Formulae Sheet*.

Question 9

Describe an experiment to measure the internal resistance of a 1.5 V battery.

In your answer include:

- **a circuit diagram**
- **the procedure taken**
- **measurements taken**
- **how the measurements are used**
- **how safety is managed**

(7 marks) (QWC 2 marks)

In most papers there will be a question based on an experiment specified in your specification. There are currently six in AS1. Generally there will be space left for a labelled diagram. Use a ruler and pencil. Follow the specific instructions for the question, but give a clear step-by-step account of the method, stating apparatus used and all readings taken. Always address accuracy and reliability.

Answer to Question 9

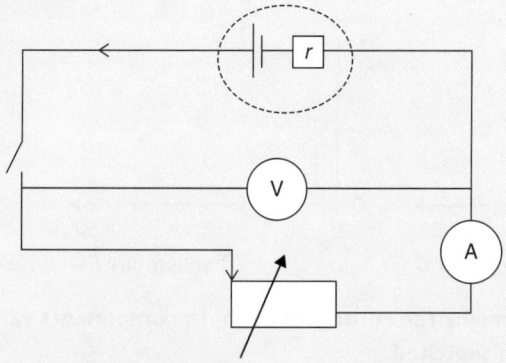

✔✔

1 Connect the circuit shown, with the variable resistor at its maximum value setting; this is the external load resistance.
2 Close the switch and immediately record the readings of current and terminal p.d. from the ammeter and voltmeter. ✔
3 Open the switch as soon as the values are recorded to reduce discharging the battery needlessly.
4 Move the slider on the variable resistor to decrease the load resistance, and repeat steps 2 and 3.
5 Repeat the procedure of reducing the load resistance several times until its value is near zero. ✔
It is important to minimise the time for which the switch is closed when the load is near zero as the current will be high and the battery could be damaged. ✔
6 Plot a graph of p.d. on the *y*-axis against current on the *x*-axis. ✔
7 Measure the *magnitude* of the slope of the graph, which is equal to the internal resistance of the battery.

$V_{\text{terminal}} = -Ir + \varepsilon$ and $y = mx + c$

gradient, $m = -r$, internal resistance ✔ (QWC) ✔✔

Quality of written communication will be assessed in one question, or part question, within the paper. This will be stated on the front cover of the paper. It will direct you to use continuous prose; when appropriate, a series of complete sentences is acceptable, as above. Correct use of capitals and punctuation, spelling and legibility are being tested, along with the scientific relevance.

Question 10

(a) Copy the axes shown below and sketch graphs to show the variation with temperature of the resistance of:

(i) a pure metallic conductor (2 marks)

(ii) a negative temperature coefficient (ntc) thermistor (2 marks)

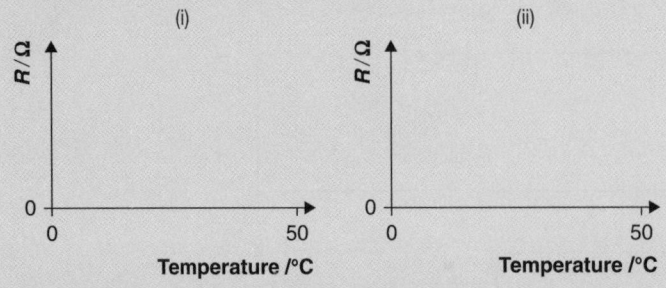

(iii) In terms of molecular theory, explain why the resistances of both components vary with temperature in the manner you have sketched. (4 marks)

(b) Superconductivity is a phenomenon observed in several metals and ceramic materials. With the help of a sketch graph, explain what is meant by superconductivity. Label the axes with both quantities and appropriate units. Also label the significant feature of the plot. (6 marks)

Answer to Question 10

(a)

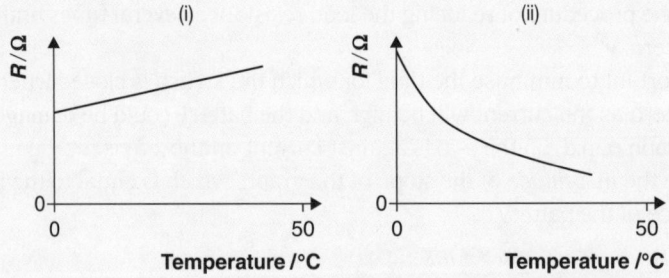

(i) non-zero start ✔; increasing linear or slight curve ✔

(ii) non-zero start ✔; decreasing curve but not cutting axis ✔

(iii) For the metal: as the temperature increases, the positive ion vibration is greater and the interaction between electrons and ions is more substantial — hence greater resistance. ✔✔

For the thermistor: this mechanism also occurs but is more than offset by the increase in the number of conducting electrons as the temperature increases — so resistance decreases. ✔✔

(b) Condition displayed in some materials of an *abrupt* drop to zero resistance as they are cooled to temperatures ranging from near absolute zero (0 K) to about 100 K. ✔✔

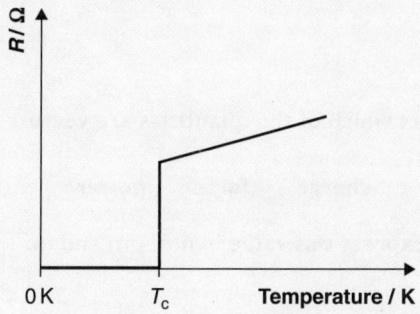

T_c = transition or critical temperature

Axes: R/Ω ✔; T/K ✔

Shape of sketch ✔

Transition labelled ✔

Total: 75 marks

Self-assessment test 2

Question 1

(a) Six physical quantities are listed below. Select which of the quantities are vectors and state their SI base units. (4 marks)

kinetic energy frequency velocity charge force power

(b) The radius of the hydrogen nucleus is 1 fm. Express this value in nm, μm and m. (3 marks)

Answer to Question 1

(a) velocity, ✔ $m\,s^{-1}$; ✔ force, ✔ $kg\,m\,s^{-2}$ ✔
(b) 1×10^{-6} nm ✔, 1×10^{-9} μm ✔, 1×10^{-15} m ✔

Question 2

The graph below shows the variation in the displacement *s* with time *t* for a car of mass 1200 kg, which is travelling in a straight line. After *t* = 2 s, the car's brakes are applied and the car decelerates uniformly to rest, stopping at time *t* = 5 s.

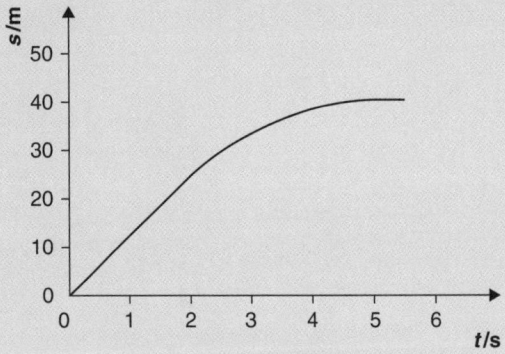

Determine each of the following quantities, showing clearly in each case how you obtain your answer:

(a) the speed of the car at time *t* = 2 s (1 mark)

(b) the value of the deceleration (2 marks)

(c) the force required to produce the deceleration (2 marks)

(d) the distance travelled by the car while slowing down (1 mark)

(e) the work done in stopping the car (2 marks)

Answer to Question 2

(a) straight line up from $t = 2\,s$

The slope of the displacement–time graph equals the velocity, which is the same as the speed because it is in a straight line up to this point.

$$\text{speed} = \frac{\Delta s}{\Delta t} = \frac{24}{2} = 12\,m\,s^{-1} \checkmark$$

(b) $a = \dfrac{\Delta v}{\Delta t} = \dfrac{(v - u)}{t} \checkmark$

$$a = \frac{(0 - 12)}{(5 - 2)} = -4\,m\,s^{-2}$$

deceleration = $4\,m\,s^{-2}$ ✔

The negative should be dropped from the answer line as the question has not asked for acceleration but deceleration — the word incorporates the negative.

(c) $F_{resultant} = ma$ ✔

$F_{resultant} = 1200 \times (-4) = -4.8\,kN$ (the negative indicates direction) ✔

(d) From the graph we can read off the displacement values at $t = 2\,s$ and $t = 5\,s$.

$\Delta s = 42 - 24 = 18\,m$ ✔

(e) work done = force × distance moved in the direction of the force ✔

work done = $-4800 \times 18 = (-)86\,400\,J$ ✔

The negative value for work is a consequence of the direction of displacement and the force being opposite. As work is a scalar, direction is irrelevant and we deal with only its magnitude. In this case work has been done *on* the force, rather than *by* the force.

Question 3

(a) A car of mass 950 kg is moving up a road inclined at 12° to the horizontal with an acceleration of $0.50\,m\,s^{-2}$. At an instant its speed is such that the opposing force of friction experienced by the car is 510 N. Calculate the driving force when the car is moving at this speed. (5 marks)

(b) The maximum speed of the car up this slope is $18\,m\,s^{-1}$. When moving at this speed the opposing force of friction has increased to 650 N.

(i) Calculate the driving force when moving at the constant speed of $18\,m\,s^{-1}$. (3 marks)

(ii) What is the mechanical power produced by the car at this speed? (2 marks)

(iii) If at this speed the car engine uses fuel energy at a rate of 180 kW, determine the percentage efficiency of the engine at this speed. (2 marks)

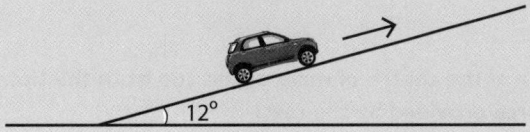

Answer to Question 3

(a) force down the incline = $F_{friction} + mg\sin\theta$ ✔

$F_{resultant} = F_{driving} - (F_{friction} + mg\sin\theta)$ ✔

$F_{resultant} = ma$ ✔

$F_{driving} = (950 \times 0.50) + (510 + [950 \times 9.81 \sin 12°])$ ✔

$F_{driving} = 2900\,N$ (to 2 s.f.) ✔

(b) (i) At maximum speed there is no acceleration. ✔

$F_{resultant} = 0$; $F_{driving} = F_{friction} + mg\sin\theta$ ✔

$F_{driving} = 650 + (950 \times 9.81 \sin 12°)$

$F_{driving} = 2600\,N$ (to 2 s.f.) ✔

The answer is given to 2 significant figures as the data provided are to 2 significant figures.

(ii) $P = Fv$ ✔

$P = 2600 \times 18 = 47\,kW$ ✔

(iii) efficiency = $\frac{\text{power output}}{\text{power input}}$ ✔

efficiency = $\frac{47}{180} = 0.26$

percentage efficiency = $0.26 \times 100 = 26\%$ ✔

Question 4

(a) Define the moment of a force about a point. (1 mark)

(b) State the principle of moments. (3 marks)

(c) When playing long shots snooker players often use a cue support known as a rest. The snooker cue itself is a non-uniform rod of length 150 cm and mass 0.57 kg. If the cue is placed on the rest 40 cm from the tip, as shown in the diagram below, the player has to apply a force of 3.0 N at the other end to balance the cue.

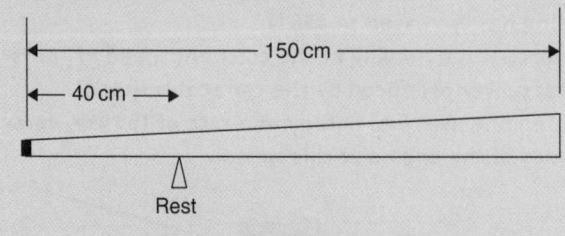

(i) Determine the distance of the centre of mass of the cue from the tip. (4 marks)

(ii) What is the support force provided by the rest? (2 marks)

Answer to Question 4

(a) The moment of a force about a point is defined as the product of the force and the *perpendicular* distance of the line of action of the force from the point. ✔

(b) When an object is in rotational equilibrium ✔, the sum of the clockwise moments about a point is equal to the sum of the anticlockwise moments ✔ about the same point ✔.

Be careful to include all the points in definitions, to access all the marks allocated. This can vary from paper to paper — for example, the definition in (a) could have been allocated another mark for 'perpendicular'.

(c)

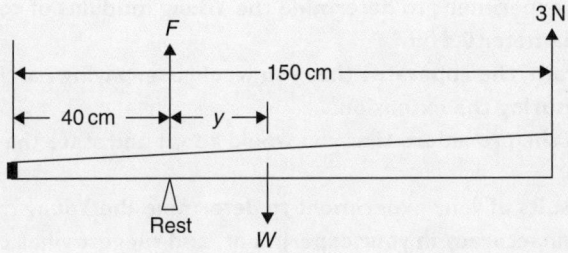

✔

With moments questions it is essential to have a good labelled diagram.

(i) Take moments about the 'rest' (pivot).

Let the distance to the centre of gravity be 'y':

clockwise moments = $W \times y = (0.57 \times 9.81) \times y$

anticlockwise moments = $3 \times (150 - 40)$

For rotational equilibrium:

$(0.57 \times 9.81) \times y = 3 \times (150 - 40)$ ✔

$y = 59\,\text{cm}$ ✔

So the distance from the tip to the centre of gravity = 99 cm. ✔

Note the use of centimetres rather than converting to metres. This is an acceptable sub-multiple, but must be used consistently throughout the question. Although the answer is technically given to 3 significant figures, because the addition takes us beyond 100, it makes sense here as the data given were to the nearest centimetre.

(ii) For vertical equilibrium:

$F_{\text{resultant}} = 0$, or $F_{\text{up}} = F_{\text{down}}$ ✔

Therefore:

$3 + F = W = (0.57 \times 9.81)$

$F = 2.6\,\text{N}$ ✔

Question 5

(a) The Young modulus for copper is 1.2×10^{11} Pa.
What is the Young modulus a measure of and how is it defined? (2 marks)

(b) The yield stress value for copper is 7.5×10^{7} Pa. If the stress is increased beyond this value there is a marked increase in the extension and the internal structure of the copper changes. A tensile load of 5.0 N is applied to a 2.4 m length of copper wire of cross-sectional area 0.10 mm^2.

(i) Explain whether such a load can be applied without the wire becoming permanently extended. (2 marks)

(ii) Calculate the extension produced by the load. (3 marks)

(c) You have been asked to perform an experiment to determine the Young modulus of copper using copper wire of approximate diameter 0.4 mm.

(i) Describe, with the aid of a diagram, the apparatus that you would use, paying particular attention to the method of measuring the extension. (3 marks)

(ii) Describe the important steps in the procedure that you would adopt and state the measurements you would make. (3 marks)

(iii) Show how you would use the results of your experiment to determine the Young modulus. (4 marks)

(iv) Identify the principal sources of inaccuracy in your experiment, and suggest what could be done to reduce these inaccuracies. (3 marks)

(QWC 2 marks)

Answer to Question 5

(a) It is a measure of the opposition of a material to deformation. ✔
It is defined as the ratio of stress/strain. ✔

(b) (i) $\text{stress} = \dfrac{\text{force}}{\text{area}} = \dfrac{5.0}{(0.10 \times 10^{-6})} = 5.0 \times 10^{7}\,\text{Pa}$ ✔

This value is less than the yield stress value, so the conclusion is that the deformation is not permanent, but elastic. ✔

(ii) $E = \dfrac{\text{stress}}{\text{strain}} = \dfrac{(F/A)}{(x/L)}$ ✔

Rearranged:

$x = \dfrac{FL}{AE} = \dfrac{F}{A} \times \dfrac{L}{E}$ ✔

$x = 5.0 \times 10^{7} \times (2.4/(1.2 \times 10^{11})) = 1.0 \times 10^{-3}\,\text{m}$ or $1\,\text{mm}$ ✔

ⓔ It is easy to lose a mark through a 10^{n} error. Use the full space available, rearrange your equation first and record each value clearly, rather than doing it in your head and entering values directly into the calculator.

(c) (i)

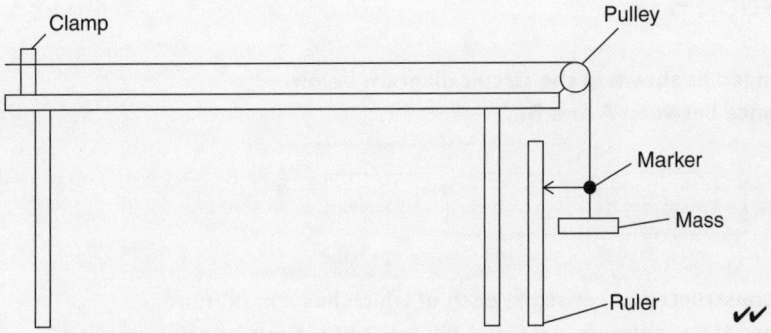

✔✔

- Clamp at least 2 metres of the wire and pass it over a pulley.
- Adjust the vertical ruler or cut the wire so that the zero of the scale is level with the end of the unloaded wire.
- Hang a 100 g slotted mass from the end of the wire, equivalent to a load of 0.98 N, and record the new position of the end of the wire on the ruler — this is the extension.
- Repeat this process several times, with each added 100 g providing a tensile force of 0.98 N. ✔

(ii)
- Measure the full length, L, of the unloaded wire with a tape measure. ✔
- Measure the diameter of the wire using a micrometer screw gauge. ✔
- Record the load attached, F, and the extension, x, produced for a range of loads. ✔

(iii) $E = \frac{\text{stress}}{\text{strain}} = \frac{(F/A)}{(x/L)}$ (already credited)

Rearranged with F as the subject:

$F = (EA/L)x$

$y = mx$

Plot a graph of load against extension, F against x. ✔

Calculate the slope, $m = \frac{EA}{L}$ ✔

Calculate A, using $A = \frac{\pi d^2}{4}$ ✔

Calculate E, using $E = \text{slope} \times (L/A)$ ✔

(iv)
- Diameter value, double uncertainty as 'd^2'; measure it in three places and take the average. ✔
- Initial length; tension the wire before use to remove the kinks. ✔
- Extension; repeat the readings as you remove the slotted masses, assuming the elastic limit has not been passed. ✔

(QWC) ✔✔

ⓔ Quality of written communication will be assessed in one question, or part question, within the paper. This will be stated on the front cover of the paper. It will direct you to use continuous prose; when appropriate, a series of complete sentences is acceptable, as above. Correct use of capitals and punctuation, spelling and legibility are being tested, along with the scientific relevance.

Question 6

(a) Three 6 Ω resistors are arranged as shown in the circuit diagram below. What is the effective resistance between A and B? (4 marks)

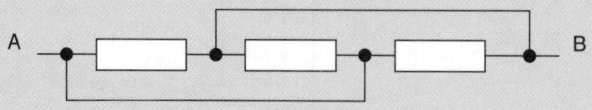

(b) The circuit shown below is constructed of resistors, each of which has a maximum power rating of 1 W. Calculate the maximum potential difference that can be safely applied across AB without damage to any resistor. You must show your working clearly. (6 marks)

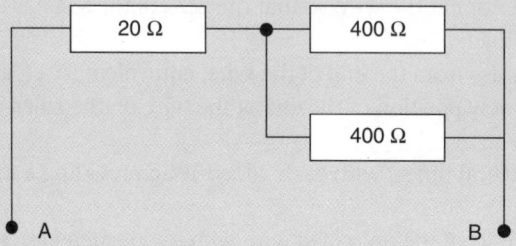

Answer to Question 6

(a)

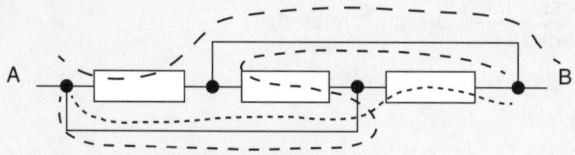

Identify two different routes for the current. ✔
Identify a third different route for the current. ✔
Effectively three 6 Ω resistors in parallel. ✔

$$\frac{1}{R_{\text{effective}}} = \frac{1}{R_1} + \frac{1}{R_2} + \frac{1}{R_3} = \frac{3}{6}$$

$R_{\text{effective}} = 2\,\Omega$ ✔

(b) Need to identify which resistor would reach the maximum power first.
Recognise the current, I, splitting equally through the 400 Ω resistors. ✔

Using $P = VI = I^2R = \frac{V^2}{R}$: ✔

for 20 Ω resistor, $P = I^2 \times 20$

for 400 Ω resistor, $P = \left(\frac{I}{2}\right)^2 \times 400 = 100 \times I^2$

So the power usage is 5 times greater in the 400 Ω resistors, so we must set the safe p.d. based on their limits. ✔

$P = 1\text{ W}$; $R = 400\,\Omega$

$P = \frac{V^2}{R}$

$V^2 = PR = 1 \times 400$

Therefore:

$V_{400} = 20\,V$ ✔

Effective resistance of the two 400 Ω resistors in parallel is 200 Ω. ✔
Therefore the p.d. drop across the 20 Ω is 20/10 = 2 V.
The maximum p.d. that can be applied safely is 22 V. ✔

Both parts of this difficult question would be classed as unstructured, so you must show each stage of your attempted solution and the relevant equations. The equations and components of the solution will draw independent marks.

Question 7

(a) The diagram below shows three wires meeting at a junction in a circuit.

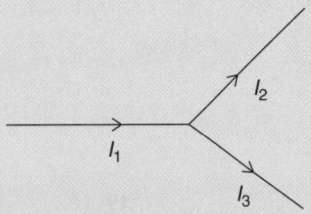

(i) The wires carry currents I_1, I_2 and I_3, as shown. Write down the equation relating all three currents.

(ii) If the values of I_1 and I_2 are both 3 A, what is the value of I_3? (2 marks)

(b) The sensor circuit shown below provides a voltage *V* that depends on the brightness of the lighting in a room.

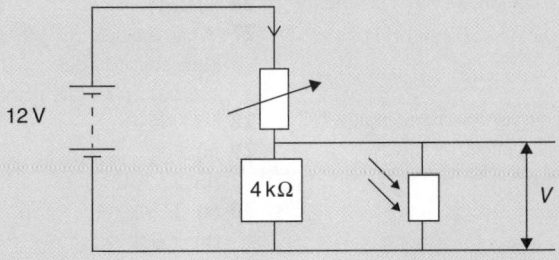

The battery has an e.m.f. of 12 V and negligible internal resistance.

The light sensor is a light-dependent resistor (LDR). The resistance of the LDR is 5.0 kΩ in the dark and 0.20 kΩ when the natural light in the room is at its brightest.

A voltage, *V* = 3 V is required to trigger the lighting circuit.

The variable resistor *R* acts as a sensitivity control for the circuit.

Find the value of *R* that will cause the lighting circuit to activate when the LDR has a resistance of 4 kΩ, just when it is getting dark. (4 marks)

Answer to Question 7

(a) **(i)** $I_1 = I_2 + I_3$ ✔

(ii) $3 = 3 + I_3$

$I_3 = 0\,A$ ✔

(b)

$$\frac{1}{R_{effective}} = \frac{1}{R_1} + \frac{1}{R_2}$$ ✔

$$\frac{1}{R_{effective}} = \frac{1}{4} + \frac{1}{4}$$

$R_{effective} = 2\,k\Omega$ ✔

Amended version of potential divider equation:

$$V_{out} = \frac{R_{effective}V_{in}}{(R_{effective} + R)}$$ ✔

$3 = (2 \times 12)/(2 + R)$

$(2 + R) = (2 \times 12)/3$

$R = 6\,k\Omega$ ✔

Total: 75 marks

Knowledge check answers

1 A unit from which other quantities can be derived.
2 s, m, kg, A, K, mol, (cd)
3 2.4×10^{-8} m
4 work, scalar; density, scalar; charge, scalar; momentum, vector
5 Scalar: speed, time, distance, gravitational potential energy, kinetic energy
Vector: velocity, displacement, acceleration, weight, momentum
6 **(a)** 7 N
(b) 17 N
(c) 13 N
7 Pushing is more effective because the vertical component will act downward to add to the flattening effect.
8 horizontal: $10.4\,m\,s^{-1}$, vertical: $6.0\,m\,s^{-1}$
9 50 m
10 $5.9\,m\,s^{-2}$
11 12.5 m
12 120 m, $2.5\,m\,s^{-2}$
13 5.7 s
14 Parabola
15 **(a)** $a = -g$, $u_v = 5.0\,m\,s^{-1}$, $u_h = 8.7\,m\,s^{-1}$
(b) $a = -g$, $u_v = 0\,m\,s^{-1}$, $u_h = 8.7\,m\,s^{-1}$
16 3.4 s (symmetry)
17 9810 N
18 600 N
19 440 N
20 **(a)** 40 N
(b) 20 N
21 87.2 N
22 2.3×10^5 J
23 $6.3\,m\,s^{-1}$
24 24000 J
25 75%
26 $80\,N\,m^{-1}$
27 When the deforming force is removed from an object that has been elastically deformed it will return to its original shape and size, but if plastically deformed it will not.
28 1.3×10^9 Pa
29 **(a)** 90 C
(b) 5.6×10^{20}
30 **(a)** 12.5 A
(b) 2.7 MJ
31 4.0 Ω
32 96 J
33 **(a)** 36 Ω
(b) 4 Ω
34 $\rho = 1.1 \times 10^{-6}\,\Omega\,m$
35 **(a)** When delivering current.
(b) When no current being drawn.
(c) When being charged by an external source.

Note: **bold** page numbers indicate definitions of key terms.

A

abbreviations 57
acceleration, definition 13
acceleration of free fall 16–17, 19, 24

C

centre of mass/gravity 26–28
charge, conservation of 39–40
circuits
  conservation laws 39–40
  questions & answers 77–78
  series and parallel 40–43
command terms 56
conservation of energy 30–31
  applied to circuits 39–40
critical temperature 46, 69
current 37
current–voltage characteristics 43–45

D

deformation of solids 34
  Hooke's law 34–35
  Young modulus 35–36
displacement–time graphs 14–15, 70–71

E

efficiency **32–33**
elastic limit 34
electrical resistance *see* resistance
electric current **37**
electricity 37
  current 37
  electromotive force 38
  potential difference 38
  power and energy 38
  resistance 39
e.m.f. (electromotive force) **38**, 40, 50–52
energy **30**
  conservation of 30–31, 40
  and electrical power 38
  transfer 30
equations
  of motion 14
  and units 7

F

forces
  electromotive 38
  frictional 23–24
  moment of a force 25
  questions & answers 71–73
  and work done 29–30
formulae sheet for AS1 exam paper 57
free-fall acceleration 16–17, 19, 24
frictional forces 23–24, 63

G

graphs
  current–voltage 44–45
  displacement–time 14–15
  resistance–temperature 46
  velocity–time 15–16
gravitational potential energy **30**
gravity
  acceleration due to 16–17
  centre of 26–27, 73

H

heater circuit 55
homogeneous (balanced) equations 7
Hooke's law **34–35**

I

internal resistance 50–51
  experiment to determine 51–52

J

joule **29**

K

kinetic energy 30
Kirchhoff's laws 39–40

L

LDRs (light-dependent resistors) 55, 77
lighting circuit 55
linear motion 12
  acceleration of free fall 16–17
  definitions of quantities 13–14
  displacement–time graphs 14–15
  equations of motion 14
  velocity–time graphs 15–16

M

moment of a force **25**
motion
  equations of 14
  linear 12–14
  Newton's laws of 22–24
  projectile 20–22

N

Newton's laws of motion 22–24

O

Ohm's law **45**
  resistance and temperature 45–46
  superconductivity 46–47

P

parabolic path, projectiles 20
parallel circuits 40–41
physical quantities 6
  equations and units 7–8
  prefixes for SI units 7
  SI base units 6
  SI derived units 6–7

plasticity 34
potential difference **38**
potential divider 52
  loading 53–54
  sensors 55
  variable potential 53
potential energy, gravitational 30
power **32**
principle of the conservation of energy **30**
principle of moments **25–26**
  centre of mass/gravity 26–28
  vertical equilibrium 28–29
projectile motion **20–22**
Pythagoras' theorem 10

Q

quantities
  linear motion 12–13
  physical 6–8
  scalars and vectors 9–12
  symbols for 6, 7, 57

R

ratios
  efficiency 32
  Young modulus 35
resistance (electrical) **39**
  of a thermistor 45
  factors determining 47–48
  internal 50–52
  parallel circuit 41
  questions & answers 64–69, 76–78
  series circuit 40
  and temperature 45–46
  zero, superconductors 46
resistivity **47–48**
  experiment to determine 48–49
resultant (sum of two vectors) 9
revision tips 57

S

scalars and vectors 9
  adding 9–10
  resolution of vectors 10–12
scale drawings 10
sensors 55, 77
series circuits 40
SI units
  base units 6
  derived units 6–7
  and equations 7
  prefixes and symbols 7, 57
solids, deformation of 34–36
speed, definition 13
strain **35**
stress **35**
superconductivity 46–47
'suvat' equations 14
symbols for quantities 6, 7, 12, 57

T

temperature
  and resistance 44–46
  and superconductivity 46–47
terminal potential difference 50
thermistors 45, 55, 68
transition (critical) temperature 46
trigonometry 10

U

ultimate tensile stress 36
units *see* SI units

V

variable potential 53
vectors and scalars 9
  addition of 9–10
  resolution of 10–12
velocity, definition 13
velocity–time graphs 15–16
vertical equilibrium 28–29
$V_{terminal}$ **50–51**

W

weight
  and moment of a force 25–29
  and Newton's second law 24
work done 29–30

Y

Young modulus 35–36, 74–75